The Rise and Fall of the Fifth Force

Allan Franklin • Ephraim Fischbach

The Rise and Fall of the Fifth Force

Discovery, Pursuit, and Justification
in Modern Physics

Second Edition

Allan Franklin
Department of Physics
University of Colorado
Boulder
Colorado, USA

Ephraim Fischbach
Department of Physics and Astronomy
Purdue University
West Lafayette
Indiana, USA

ISBN 978-3-319-28411-8   ISBN 978-3-319-28412-5   (eBook)
DOI 10.1007/978-3-319-28412-5

Library of Congress Control Number: 2016933674

Springer Cham Heidelberg New York Dordrecht London
© Springer International Publishing Switzerland 2016

Printed on acid-free paper

Springer International Publishing AG Switzerland is part of Springer Science+Business Media (www.springer.com)

Preface to the Second Edition

It might seem odd to publish a second edition of a book that deals with an episode in science, the existence of a Fifth Force, that was presumably decided more than 20 years ago. The primary reason for this is the availability of a first-person account of the origin of that hypothesis by one of its proposers. We believe that such an account provides interesting and important insights into the practice of physics, which are not usually available in standard historical accounts. This also provides us with an opportunity to discuss the idea that wrong science is not necessarily bad science. As the subsequent history shows, that hypothesis led to interesting and important experiment and theoretical work, even after its supposed demise.

Boulder, USA   Allan Franklin
West Lafayette, USA   Ephraim Fischbach
October 2015

Preface to the First Edition

On January 8, 1986, *The New York Times* announced, "Hints of Fifth Force in Nature Challenge Galileo's Findings." In January 1990, at an informal discussion meeting, which included many of those working on the Fifth Force, Orrin Fackler stated, "The Fifth Force is dead." No one present disagreed.

In this essay I will examine the short, happy life of the Fifth Force hypothesis.[1] This study will allow us to examine the roles that evidence plays in the contexts of discovery, of pursuit, and of justification.[2] We will be able to look at two of the interesting questions in the history and philosophy of science. How is a hypothesis proposed, and how and why does it become considered worthy of further theoretical and experimental investigation by the scientific community? These are the contexts of discovery and of pursuit.

It may well be that the suggestion of hypotheses or theories is the free creation of an individual scientist, but I doubt that such creative events occur in a vacuum. Thus, although Newton's thoughts on the universality of gravitation may have been triggered by the apple falling on his head, it seems unlikely that it would have had that effect had he not already been thinking about gravitation and the motion of the moon.[3] I suggest that when scientists offer hypotheses, they, or the rest of the scientific community, may have been considering the problem for a time. In addition to solving the problem, the hypothesis is also likely to be supported by some existing empirical evidence or has some theoretical plausibility because

---

[1]For a technical discussion of the Fifth Force, see Fischbach and Talmadge (1992). A complete bibliography of papers on non-Newtonian gravity, including the Fifth Force, is given in Fischbach et al. (1992).

[2]Although the discussion of these contexts is regarded as somewhat unfashionable at the moment, I believe that this study will show that it can provide a useful framework for approaching not only this particular episode but for the history of modern physics in general.

[3]Professor Sam Westfall, the noted Newton scholar and biographer, believes the story of Newton and the apple is true. He reports that Newton repeated it on at least four occasions during his lifetime (private communication).

it resembles previous successful solutions of other problems.[4] The scientist may also consider the interest of and importance of the hypothesis or theory. One might also require, for both proposing and pursuing, that the hypothesis be consistent with existing, well-confirmed theories, as well as with existing empirical evidence.[5] It may also fit in with an existing research program or look like a fruitful or interesting line of research. Another factor may be that the theory has desirable mathematical properties. Thus, for example, the Weinberg–Salam unified theory of electroweak interactions did not receive much attention until t'Hooft showed it was renormalizable in 1971.[6] These are also reasons why a theorist may pursue a hypothesis.

Experimentalists planning to investigate a hypothesis may also have similar reasons for their work. In addition, there may be what one might call experimental reasons for such pursuit. These may include the fact that the proposed measurement can be done with existing apparatus or with small modifications of it. The measurement may fit in with an existing series of measurements in which the experimenter(s) has expertise or the experimenter may think of a clever way to perform the measurement. If the hypothesis is sufficiently important, the experimenter may even construct an entirely new apparatus. At this point the cost of the experiment, the availability of research funds, as well as the perceived interest and importance

---

[4]I do not suggest that these are necessary or sufficient conditions for such hypothesis creation, after all an individual might come up with a solution on the first try, but my speculation is that these are the usual circumstances.

[5]One should not try to make this concept of consistency too rigorous. One shouldn't, and in fact one can't, require formal consistency when the theory proposed is being offered as an alternative to an existing, well-confirmed theory. For example, Einstein's Special Theory of Relativity, which required that the mass of an object be a function of its velocity, was inconsistent with Newtonian mechanics, in which mass is a constant. There was, however, a kind of consistency at the empirical level. In the region in which velocities were small compared to c, the speed of light, and in which Newtonian mechanics had been well-confirmed, the predictions of both theories agreed within the experimental uncertainties of the results. It was only at higher velocities that the theories gave very different results, and these had not been tested. Thus, one might require that the two theories agree with both the data and each other to within the stated experimental uncertainties. Even here, I believe, one shouldn't be too strict. After all, Newton's mechanics which showed that Kepler's law that the planets move in ellipses with the Sun at one focus was incorrect because of the perturbations due to other planets. An examination of the data at the time indicated that there were such deviations, although they were quite small. The new hypothesis might cause one to reexamine the evidence. Similarly, in the case of parity conservation (Franklin 1986, Chap. 1), Lee and Yang reexamined the evidence supporting the hypothesis and found, to their surprise, that the hypothesis had been tested only in the strong and electromagnetic interactions and not in the weak interactions. The evidence was not, in fact, what the physics community had thought it was. I believe, however, that consistency with accepted theory and evidence, in some sort of rough and ready way, is a criterion for pursuit. One should allow some leeway for the new theory, however. As will be discussed later, scientists do not even have to believe in the truth of a hypothesis, or in the evidence supporting it, in order to pursue that hypothesis.

[6]Pickering (1984, p. 106) noted that the citation history of Weinberg's paper clearly shows this. T'Hooft showed the theory was renormalizable in 1971. The citations were 1967, 0; 1968, 0; 1969, 0; 1970, 1; 1971, 4; 1972, 64; and 1973, 162.

of the experiment and hypothesis will certainly enter into the decision to do the experiment, but that is left for future discussion.

As we shall see below, the suggestion of a "Fifth Force" in gravitation occurred after the authors had been worrying about the problem for some time, did have some empirical support, and also resembled, at least in mathematical form, Yukawa's previous successful suggestion of the pion to explain the nuclear force. It also fit in with the previous work on modifications of gravitational theory by Fujii and others.

A related question is how decisions concerning the fate of such hypotheses are made. Is their confirmation or refutation, based primarily on valid experimental evidence, as I have previously argued (Franklin 1990), or do other considerations enter? This is the context of justification.

In this essay I will consider the history of the Fifth Force in gravitation and offer tentative and partial answers to the questions of hypothesis creation and pursuit.[7] This history will offer a further illustration that evidence is decisive and crucial in the context of justification. In this episode I will look at how and why Ephraim Fischbach, Sam Aronson, Carrick Talmadge, and their collaborators came to suggest modifying gravitational theory by adding such a force.[8] I will also examine the evidential context at the time that led at least a segment of the physics community to investigate this hypothesis. I will also discuss the subsequent history of this hypothesis, up to the present. At the present time, I believe it is fair to say that the majority of the physics community does not believe that such a force exists. The current experimental limits on the strength of such a force are approximately 10^{-4} that of the normal gravitational force (this depends on the choice of coupling, i.e., baryon, isospin, etc., and on the assumed range of the force). There is also good evidence that the distance dependence of the gravitational force is $1/r^2$. Although, as discussed below, some experimental anomalies remain, they are not presently regarded as serious. There are, in addition, other experimental results which contradict the anomalous results, and the overwhelming preponderance of evidence is against the existence of the Fifth Force.

---

[7]These answers are necessarily partial because a complete answer would need a worked out theory of plausibility, how scientists decide which hypotheses should be included in the space of plausible and interesting conjectures. This is an extremely important issue in both the contexts of pursuit and discovery. In this particular case there were only two competing theories, (1) existing gravitational theory and (2) gravitational theory plus the Fifth Force, so the issue of plausibility is simplified. I will argue below that evidence made the second alternative worth pursuing, at least for a segment of the physics community.

[8]These three physicists played the leading roles in the formulation of the Fifth Force hypothesis. I will refer to papers written by them and their collaborators by the first author listed. This may give the impression that only a single author was involved. This is definitely not the case. Virtually all of these papers had multiple authors.

Acknowledgments

One of the advantages of working on the history of a contemporary episode is that you can speak with the participants. Ephraim Fischbach and Sam Aronson, two of the originators of the Fifth Force hypothesis, have always been available for discussion of both technical points and of the history. Ephraim provided a preprint of his bibliography of articles on the Fifth Force, and his encouragement has been invaluable. Sam provided me with a copy of the e-mail record, which allowed a unique look into the private practice of science. Although both of them have read portions of this manuscript in various forms, neither of them has asked for any changes or modifications of the story. I take full responsibility for this text.

I have also benefitted from discussions with David Bartlett and Jim Faller, two of my colleagues at the University of Colorado, as well as with Eric Adelberger, Paul Boynton, Don Eckhardt, and Riley Newman. David made very helpful comments regarding the manuscript, and Eric, Jim, and Paul provided me with photographs of their experiments. Conversations with Mara Beller on both the historical and philosophical issues were also very helpful. I am also grateful to John Monnier and Alex Hansen for reading this manuscript and suggesting points that needed further clarification or explanation.

Part of this work was done while I was a visiting professor in the Department of History and Philosophy of Science, Indiana University, and also in the Division of Humanities and Social Sciences, California Institute of Technology. I am grateful to my colleagues at both places for stimulating discussions and to the department and the division for their hospitality and support. Part of this work was supported by a faculty fellowship and grant-in-aid from the Council on Research and Creative Work, Graduate School, University of Colorado, and I thank the Council for its support. This material is based upon work partially supported by the National Science Foundation under Grant No. DIR-9024819. Any opinions, findings, and conclusions or recommendations expressed in this material are those of the author, and do not necessarily reflect the views of the National Science Foundation.

Boulder, USA   Allan Franklin
1993

References

Fischbach, E., Talmadge, C.: The Search for Non-Newtonian Gravity. American Institute of Physics, Springer (1992)

Fischbach, E., et al.: Non-Newtonian gravity and new weak forces: an index of measurements and theory. Metrologia **29**, 213–260 (1992)

Franklin, A.: The Neglect of Experiment. Cambridge University, Cambridge (1986)

Franklin, A.: Experiment, Right or Wrong. Cambridge University, Cambridge (1990)

Pickering, A.: Against putting the phenomena first: the discovery of the weak neutral current. Stud. Hist. Philos. Sci. **15**, 85–117 (1984)

Contents

Part I
The Rise and Fall of the Fifth Force

Chapter 1
The Rise ...

1.1   K Mesons and CP Violation

The story of the Fifth Force begins with a seeming digression because it involves not a modification of gravitational theory, but rather an experimental test of and confirmation of that theory. In 1975 Colella, Overhauser, and Werner measured the quantum mechanical phase difference between two neutron beams caused by a gravitational field. Although these experiments showed the effects of gravity at the quantum level, they did not, in fact, distinguish between General Relativity and its competitors, as Fischbach pointed out (1980; Fischbach and Freeman 1979). This was because these experiments were conducted at low speeds, and in the nonrelativistic limit all existing gravitational theories, such as General Relativity and the Brans–Dicke theory, reduce to Newtonian gravitation. Fischbach also discussed how one might test general relativity at the quantum level by considering gravitational effects in hydrogen.

In this work, partly as a result of conversations with Overhauser, Fischbach went on to consider whether or not gravitational effects might explain the previously observed violation of CP symmetry (combined charge-conjugation or particle–antiparticle symmetry and parity or space reflection symmetry) in K_L^0 decays.[1] He had shown that an external gravitational field resulted in an admixture of atomic states of opposite parity. For a two-fermion system, such as positronium or charmonium, this also leads to a change in the eigenvalue of CP. This made "it natural to attempt to connect (the gravitational effect) with the known CP-violating K_L^0 decays (Fischbach 1980, p. 371)." Although, as Fischbach noted, there were both

---

[1] There are two different K^0 mesons, the short-lived K_S^0, and the longer lived K_L^0. CP symmetry allows the K_S^0, but not the K_L^0, to decay into two pions. For a detailed discussion of CP symmetry and the discovery of its violation see Franklin (1986, Chap. 3).

© Springer International Publishing Switzerland 2016
A. Franklin, E. Fischbach, *The Rise and Fall of the Fifth Force*,
DOI 10.1007/978-3-319-28412-5_1

experimental and theoretical reasons against gravity as the source of CP violation, the relevance of the arguments to his case were not clear.

The arguments Fischbach was referring to concerned attempts to explain CP violation and will be relevant to the later history as well. Bell and Perring (1964) and Bernstein et al. (1964) had speculated that a long-range external field that coupled differently to the K^0 and $\overline{K}^0$, a hyperphoton, could explain the violation. Such a field predicted that the effect would be proportional to the square of the energy of the K mesons. Weinberg (1964) pointed out that because neither strangeness nor isotopic spin, the supposed origins of the field, were absolutely conserved, the hyperphoton must have a finite mass, related to the range of the interaction. Assuming that the range of the interaction was the size of our galaxy, he calculated the ratio $(K_S^0 \rightarrow 2\pi + \text{hyperphoton})/(K_S^0 \rightarrow 2\pi)$ as 10^{19}. This implied that the K meson and all strange particles would be totally unstable, in obvious disagreement with experiment. He could explain the observations if he assumed that the range of the interaction was the size of the Earth, which he regarded as implausible. The issue became moot when the experiments of Galbraith et al. (1965) and of DeBouard et al. (1965) at very different energies from both each other and from the original experiment of Christenson et al. (1964) failed to show the predicted energy-squared dependence. In fact, the experiments indicated that the CP violation was constant as a function of energy for the energy range 1–10 GeV.

Fischbach was also motivated by what he took to be a "remarkable numerical relation". Using his calculated energy scale for the gravitational effect,[2] Δm, the known $K_L - K_S$ mass difference, and an enhancement factor of $m_K/\Delta m$, for which no justification was given, he found that his calculation of the gravitational effect for CP violation in K meson decay was equal to 0.844×10^{-3}, while the CP violating parameter[3] $1/2 \text{Re}\,\epsilon$ was approximately equal to 0.82×10^{-3}. This seems indeed to be a remarkable coincidence because there is no known connection between gravity and CP violation, or any accepted explanation of CP violation itself. It is made even more remarkable when one realizes that the enhancement factor $m_K/\Delta m = 1.4 \times 10^{14}$.

Fischbach continued to work on the question of how to observe gravitational effects at the quantum level. A relativistic version of the Colella, Overhauser, and Werner experiment using neutrons did not seem feasible, so he turned his attention to K mesons, where such experiments did seem possible. He began a collaboration with Sam Aronson, an experimental physicist with considerable experience in K meson experiments. At this time Aronson and his collaborators had been investigating the regeneration of K_S^0 mesons (Roehrig et al. 1977; Bock et al. 1979).[4] Although the published papers stated that (Bock et al. 1979, pp. 351–352)

---

[2]The scale of the gravitational effect was given by $g\hbar/c$, where g is the local acceleration due to gravity, $\hbar$ is Planck's constant/2π, and c is the speed of light. At the surface of the Earth this quantity is 2.2×10^{-23} eV.

[3]Here ϵ, which is a complex number, is a measure of the decay rate of K_L^0 mesons into two pions.

[4]The phenomenon of regeneration was one of the unusual properties of the K^0 mesons. If one produced these mesons in an accelerator, one obtained a beam that was 50 % K_S^0 and 50 % K_L^0.

Fig. 1.1 The phase of the
regeneration amplitude as a
function of momentum (From
Bock et al. 1979)

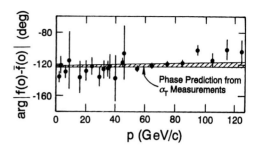

"the data are consistent with a constant phase [of the regeneration amplitude],"[5] Aronson and Bock, two members of the group, were troubled by what seemed to be an energy dependence of the phase. In fact, Bock had investigated whether changes in the acceptance could account for the effect. They couldn't. The data are shown in Fig. 1.1, along with the constant phase prediction. Although the data are consistent with a constant phase, there is at least a suggestion of an energy dependence. The low energy points have a larger phase than the high energy points.

Aronson and Bock then asked Fischbach if there was a theoretical explanation of the effect. Fischbach had none to offer. This suggested energy dependence led them to examine the possible energy dependence of the parameters of the K^0–$\overline{K}^0$ system in some detail. They found suggestive evidence for such a dependence (Aronson et al. 1982). They examined Δm, the $K_L - K_S$ mass difference, τ_S, the lifetime of the short-lived K meson, $|\eta_{+-}|$, the magnitude of the CP-violating amplitude, and $\tan \phi_{+-}$, the tangent of the phase of the CP-violating amplitude. They fitted these parameters to an energy dependence of the form $x = x_0(1 + b_x \gamma^N)$, $N = 1, 2$ and $\gamma = E_K/M_K$. They found that the coefficients differed from zero by 3, 2, 2, and 3 standard deviations, for the quantities noted above, respectively. One of their fits is shown in Fig. 1.2.

The fit to ϕ_{+-} depended on the value one assumed for ϕ_{21}, the phase of the regeneration amplitude. (Recall that it was the suggestion of an energy dependence in this quantity that led to this investigation.) The measured quantity, in fact, depends on $\phi_{21} - \phi_{+-}$. One could attribute the energy dependence to either one of them separately, or to both of them. All theoretical models at the time (see Aronson et al. (1983b) for details) predicted that over this energy range, 30–110 GeV, the change in ϕ_{21} would be less than 2°. The observed change of approximately 20° was then attributed to an energy dependence in ϕ_{+-}.

The group continued their study and presented a more detailed analysis that included data from other experiments. The most significant energy dependence,

---

If one allowed all of the K_S^0 mesons to decay and allowed the remaining K_L^0 mesons to interact with matter, one found that the beam once again contained K_S^0 mesons. They were regenerated. See Franklin (1986, Chap. 3) for details.

[5] A stronger statement had appeared earlier (Roehrig et al. 1977, p. 1118): "The results are clearly consistent with constant phase [...]."

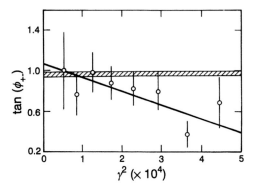

Fig. 1.2 Plot of $\tan\phi_{+-}$ as a function of γ^2 (energy squared) (From Aronson et al. (1982). The *shaded line* is the world average and the *solid line* was their fit)

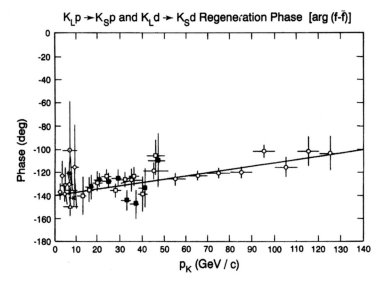

Fig. 1.3 Plot of ϕ_{+-} as a function of momentum (From Aronson et al. (1983a). See discussion in Footnote 8. For detailed references for the data see Aronson et al. 1983a)

that in ϕ_{+-}, is shown in Fig. 1.3.[6] They concluded (Aronson et al. 1983a, p. 488): "The experimental results quoted in this paper are of limited statistical significance. The evidence of a positive effect in the energy dependences of Δm, τ_S, $|\eta_{+-}|$, and ϕ_{+-} is extremely tantalizing, but not conclusive. The evidence consists of $b_x^{(N)}$'s which are different from zero by at most 3 standard deviations."

---

[6]The graph actually shows the energy dependence of ϕ_{21}, assuming ϕ_{+-} was constant. If ϕ_{21} is considered to be a constant, the graph shows the energy dependence of ϕ_{+-}.

A second paper (Aronson et al. 1983b) examined possible theoretical explanations of the effects.[7] They found (Aronson et al. 1983b, p. 495): "Using this
formalism we demonstrate that effects of the type suggested by the data [energy
dependences] cannot be ascribed to an interaction with kaons with an electromagnetic, hypercharge, or gravitational field, or to the scattering of kaons from stray
charges or cosmological neutrinos." They suggested that a tensor field mediated
by a finite mass quantum might explain the effects and concluded (Aronson et al.
1983b, p. 516): "It is clear, however, that if the data [...] are correct, then the source
of these effects will represent a new and hitherto unexplored realm of physics."[8]

The subsequent history of measurements of these quantities seems to argue
against any energy dependence.[9] Coupal et al. (1985) measured $|\eta_{+-}|$ at 65 GeV/c
and found a value $|\eta_{+-}| = (2.28 \pm 0.06) \times 10^{-3}$ in good agreement with the
low energy average (5 GeV) of $(2.274 \pm 0.022) \times 10^{-3}$, and in disagreement with
$(2.09 \pm 0.02) \times 10^{-3}$ obtained by Aronson et al. (1982).[10] Grossman et al. (1987)
measured τ_S, the K_S lifetime, over a range 100–350 GeV/c. Their results, along with
those of Aronson et al. (1982), are given in Fig. 1.4. The fits obtained by Aronson
for possible energy dependence are also shown. They concluded (Grossman et al.

Fig. 1.4 Plot of τ_S, the K_S
lifetime, as a function of
momentum (From Grossman
et al. (1987). The *curves* were
the fits from Aronson et al.
(1982))

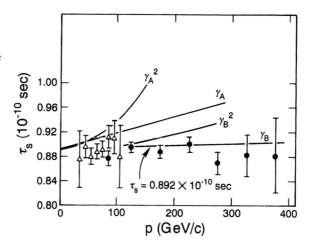

1987, p. 18): "No evidence was found for the momentum dependence suggested by the intermediate-range 'fifth-force' hypothesis." Carosi et al. (1990) measured the phases of the CP-violating amplitudes for K^0 decay in the energy range 70–170 GeV. Their results were in good agreement with those at low energy and show no evidence of any energy dependence. Still, at the time of our story, the suggested energy dependence remained a "tantalizing" effect.

1.2   Modifications of Newtonian Gravity

A second strand of our story concerns the recent history of alternatives to, or modifications of, standard gravitational theory.[11] For a time, at least, this strand was independent of the K meson story. The Fifth Force story, per se, began when the two strands were joined. Newtonian gravitational theory and its successor, Einstein's General Theory of Relativity, although strongly supported by existing experimental evidence,[12] have not been without competitors.[13] Thus, Brans and Dicke (1961) offered a scalar–tensor alternative to General Relativity. This theory contained a parameter ω, whose value determines the relative importance of the scalar field compared to the curvature of spacetime. For small values of ω the scalar field dominates, while for large values the Brans–Dicke theory is indistinguishable from General Relativity. By the end of the 1970s experiment favored a large value of ω, and thus, favored General Relativity. For example, the lunar laser-ranging experiments required $\omega > 29$, while the Viking time-delay results set a lower limit of $\omega > 500$ (Will 1981).

In the early 1970s, Fujii (1971, 1972, 1974) suggested a modification of the Brans–Dicke theory that included a massive scalar exchange particle. This was in addition to the massless scalar and tensor particles of that theory. He found that including such a particle gave rise to a force that had a short range (of the order 10 m–30 km) depending on details of the model. In Fujii's theory, the gravitational potential had the form $V = -GmM/r(1 + \alpha e^{-r/\lambda})$, where α was the strength of the

---

[11]I will be discussing here modifications of the Newtonian inverse square law, and not the well-established relativistic post-Newtonian corrections, which are of order $GM/c^2 r$.

[12]For an excellent and accessible discussion of this, see Will (1984). For more technical details, see Will (1981).

[13]The history of gravitational theory is not a string of unbroken successes. Newton himself could not explain the motion of the moon in the *Principia* and his later work on the problem, in 1694–1695, also ended in failure (Westfall 1980, pp. 442–443, 540–548). The law was also questioned during the nineteenth century when irregularities were observed in the motion of Uranus. The suggestion of a new planet by Adams and LeVerrier and the subsequent discovery of the planet Neptune turned the problem into a triumph. During the nineteenth century it was also found that the observed advance of the perihelion of Mercury did not match the predictions of Newtonian theory. This remained an anomaly for 59 years until the advent of Einstein's General Theory of Relativity, the successor to Newtonian gravitation.

interaction and λ its range. The second term was Fujii's modification. O'Hanlon (1972) suggested the same potential. This model also predicted a gravitational constant G that varied with distance.[14] Fujii calculated that the gravitational constant at large distances G_∞ would be equal to $3/4G_{LAB}$, the value at short distances.

Fujii also looked for possible experimental tests of this theory. Most interestingly for our story, he discussed the famous experimental test of Einstein's equivalence principle that had been performed by Eötvös and his collaborators (1922). (This experiment, which will be very important later in our history will be discussed below.) He noted that if his new field coupled equally to baryons (in this case the protons and neutrons in the atomic nucleus) and leptons (atomic electrons) there would be no effect, while if the field did not couple to leptons such an effect would be observed. He calculated that for an Eötvös-type experiment on gold and aluminum there would be a change in angle of 0.07×10^{-11}, for an assumed range of 40 km.[15] The best experimental limit at the time was that of Roll et al. (1964) of 3×10^{-11}, although that experiment which measured the equality of fall toward the Sun cast no light on his short range force. A note added in proof remarked that he had learned that the best estimate of the range of such a force was considerably smaller. For an assumed 1 km range he found that the change in angle was 0.5×10^{-9}. This predicted effect was smaller than the limit set by Eötvös, whose experiment was sensitive to such a short-range theory. Fujii suggested redoing the Eötvös experiment and other possible geophysics experiments, although he noted that mass inhomogeneities would present difficulties. As we shall see, Fujii's comments were prescient.

Other modifications of gravitational theory were suggested by Wagoner (1970), Zee (1979, 1980), Scherk (1979), and others. Zee's modification had a much shorter range than that of Fujii, while Scherk suggested a repulsive force with a range of about 1 km.

Long (1974) considered the question of whether or not Newtonian gravity was valid at laboratory dimensions. He found (p. 850) "that past G [the gravitational constant] measurements in the laboratory set only very loose limits on a possible variation in G and that present technology would allow a considerable improvement." He also made reference to the suggestions of Wagoner, of Fujii, and of O'Hanlon. Long (1976) proceeded to test his hypothesis experimentally, and found a small variation in G which he parameterized in the form $G(R) = G_0(1 + 0.002 \ln R)$, where R is measured in centimeters.

Long's work led Mikkelsen and Newman (1977) to investigate the status of G.[16] They used data from laboratory measurements, orbital precession, planetary

---

[14]Some readers might worry that a variable constant is an oxymoron, but it does seem to be a useful shorthand.

[15]The angle referred to is the rotation of the torsion pendulum shown in Figs. 1.6 and 1.7.

[16]The influence of Long's work is apparent in the first sentence of the abstract (Mikkelsen and Newman 1977, p. 919): "D.R. Long and others have speculated that the gravitational force between point masses in the Newtonian regime might not be exactly proportional to $1/r^2$."

mass determinations, geophysical experiments, and solar models. They concluded (p. 919): "Constraints on $G(r)$ in the intermediate distance range from $10\,\mathrm{m} < r <$ 1 km are so poor that one cannot rule out the possibility that G_c [G_∞] differs greatly from G_0 [G_{LAB}]." They pointed out (p. 924) that their analysis "does not even rule out Fujii's suggested value $G_c/G_0 = 0.75$."

The experimental study of possible violations of Newtonian gravity continued. Panov and Frontov (1979) found $G(0.3\,\mathrm{m})/G(0.4\,\mathrm{m}) = 1.003 \pm 0.006$ and $G(10\,\mathrm{m})/G(0.4\,\mathrm{m}) = 0.998 \pm 0.013$. They concluded that, despite the fact that their experimental uncertainty was larger than Long's measured effect (p. 852): "These results do not confirm the data of D.R. Long, according to which spatial variations of G do exist." Spero et al. (1980) agreed. Their measurements at distances of 2–5 cm had the required sensitivity to check Long's result and (p. 1645): "The results support an inverse-square law. Assuming a force deviating from inverse square by a factor $(1 + \epsilon \ln r)$ [Long's suggested form] it is found that $\epsilon = (1 \pm 7) \times 10^{-5}$."

Long (1981) surveyed the literature and concluded that within the quoted uncertainties all the results, including that of Panov and Frontov, were consistent with his observation. He argued that Spero's result did not, in fact, contradict his. This was because his suggested cause of the deviation, a quantum gravity vacuum polarization effect, would be significant only for a nonzero gravitational field, whereas Spero's experiment was conducted in a zero gravitational field.

The most important summary of this work, from the point of view of the subsequent history of the Fifth Force, was that given by Gibbons and Whiting (1981). Their results are shown in Fig. 1.5 for both attractive and repulsive forces. In both cases α is restricted to lie below each curve, except for curve b, which is Long's result, and in which α must lie between the two curves. They stated (p. 636): "However, conventional vacuum polarization effects in quantum gravity do not lead to the behavior required by Long: such effects are insignificant." Curve a was Spero's data, calculated at one standard deviation (s.d.); c was from Panov and Frontov; d from Mikkelsen and Newman using lunar surface gravity and Mercury and Venus flybys; e a comparison of satellite and geodesy data by Rapp (1974, 1977), the upper curve assumed agreement to 0.1 ppm (parts per million) and the lower curve 1 ppm. Rapp had reported an agreement to 2 ppm.

A different type of experiment, that of measuring gravity in either a mine (Stacey et al. 1981) or in submarines (Stacey 1978) is illustrated in curve f. The curves were *calculated*, not measured, assuming a mine experiment with an accuracy of 1 % (upper curve) and a submarine experiment with an accuracy of 0.1 % (lower curve), both for a depth of 1 km. Stacey et al. (1981) had measured G and found it to be $G = (6.71 \pm 0.13) \times 10^{-11}\,\mathrm{m^3\,kg^{-1}\,s^{-2}}$, in agreement with the laboratory value of $(6.672 \pm 0.004) \times 10^{-11}$. Their stated uncertainty included an estimate of possible systematic effects, which increased the uncertainty by about a factor of 3. They also surveyed other mine and borehole measurements of G and found them to be, in general, systematically slightly high but "tantalizingly uncertain" because of possible mass anomalies. A somewhat later paper (Stacey and Tuck 1981) gave numerical details of that survey and reported values of G, calculated in two different ways, based on a comparison of sea floor and sea surface measurements

Fig. 1.5 Plot of $\log_{10} \alpha$ vs $\log_{10}(\lambda/1$ m$)$, α is constrained to lie below the curves (From Gibbons and Whiting 1981)

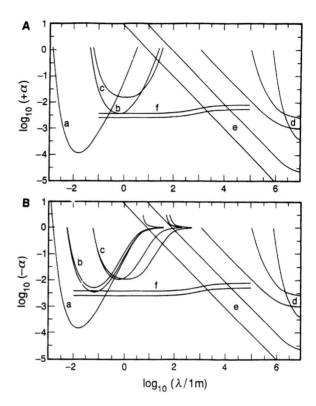

of $G = (6.730 \pm 0.010) \times 10^{-11}$ and $(6.797 \pm 0.016) \times 10^{-11}$ m^3 kg^{-1} s^{-2}, where the uncertainty is purely statistical and does not include possible systematic effects. Once again the results were higher than the laboratory value, but because of the uncertainty about possible systematic effects no firm conclusion could be drawn.

Gibbons and Whiting summarized the situation as follows (p. 636): "It has been argued that our experimental knowledge of gravitational forces between 1 m and 10 km is so poor it allows a considerable difference between the laboratory measured gravitational constant and its value on astronomical scales—an effect predicted in theories of the type alluded to above [these included Fujii and O'Hanlon, whose work was also cited in the experimental papers] [...] it can be seen that for $3\,\text{m} < \lambda < 10^3\,\text{km}$ α is very poorly constrained." Although experiment allowed for such a difference in the laboratory and astronomical values of G, there were reasonably stringent limits on any proposed modifications of the law of gravity in the distance range 1 m–10 km (p. 638): "We conclude that there is very little scope for a theory which allows deviations $>1\,\%$ from Newton's law of gravitational attraction on laboratory or larger length scales [...]. Further large scale experiments are essential to improve bounds on α between 1 m and 10 km." There was, however, a clear, although small, window of opportunity (see Fig. 1.5).

1.3   The Fifth Force

Until early 1983 the two strands, that of the energy dependence of the K^0–$\overline{K}^0$ system parameters and that of modifications of Newtonian gravity and their experimental tests, had proceeded independently. At about this time Fischbach became aware of the discrepancies between experiment and gravitational theory (the work of Stacey et al. (1981) and Stacey and Tuck (1981)).[17] He made no connection, at this time, between the two problems because he was still thinking in terms of long-range forces, which produced an energy-squared dependence of the K^0–$\overline{K}^0$ parameters, and was therefore ruled out experimentally. In early 1984, he realized that this would not be the case for a short-range force, and that the effect could be much smaller.[18] At this time he also became aware of the Gibbons and Whiting summary and realized that such a short-range force might be possible and that the two problems might have a common solution.

Fischbach, Aronson, and their collaborators looked for other places where such an effect might be seen with existing experimental sensitivity. They found only three: (1) the K^0–$\overline{K}^0$ system at high energy, which they had already studied; (2) the comparison of satellite and terrestrial determinations of g, the local gravitational acceleration[19]; and (3) the original Eötvös experiment, which measured the difference between the gravitational and inertial masses of different substances. If a short-range, composition-dependent force existed then it might show up in this experiment. They noted that the very precise modern experiments of Roll et al. (1964) and of Braginskii and Panov (1972) would not have been sensitive to such a force because they had compared the gravitational accelerations of pairs of materials toward the Sun, and thus looked at much larger distances.

The apparent energy dependence of the K^0–$\overline{K}^0$ parameters along with the discrepancy between gravitational theory and the mineshaft experiments led Fischbach, Aronson, and their colleagues (Fischbach et al. 1986) to reexamine the original data of Eötvös et al. (1922) to see if there was any evidence for a short-range, composition-dependent force.[20] By this time they knew of Holding and Tuck's (1984) result which gave G measured in a mine as $G = (6.730 \pm 0.003) \times 10^{-11}$ m^3 kg^{-1} s^{-2} in disagreement with the best laboratory value of $(6.6726 \pm 0.0005) \times 10^{-11}$. This result was still uncertain because of possible regional gravity anomalies. Fischbach used the modified gravitational potential $V(r) = -G_\infty m_1 m_2 / r(1 + \alpha e^{-r/\lambda})$. They noted that such a potential could explain

---

[17]Fischbach (private communication) attributes this to a conversation with Wick Haxton.

[18]Fischbach's first calculation was for a δ-function force.

[19]Rapp (1974, 1977) had already found $\Delta g/g \approx (6 \pm 10) \times 10^{-7}$. For the proposed Fifth Force parameters (see below) the predicted effect would be approximately 2×10^{-7}.

[20]Because the energy dependence of the K^0–$\overline{K}^0$ parameters might have indicated a violation of Lorentz invariance, Fischbach et al. (1985) had looked at the consequences of such a violation for the Eötvös experiment.

Fig. 1.6 The apparatus used
by Eötvös et al. (1922).
Although a platinum mass is
shown as the standard, this
was not always the case

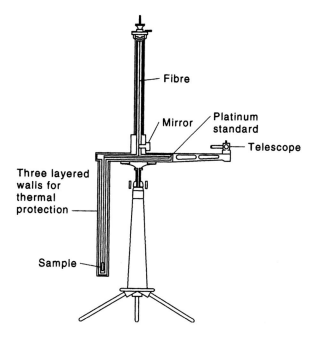

the geophysical data quantitatively if $\alpha = (-7.2\pm3.6)\times10^{-3}$, with $\lambda = 200\pm50$ m.
(This was from a private communication from Stacey. Details appeared later in
Holding et al. 1986.) This potential had the same mathematical form as that
suggested much earlier by Fujii. Recall also that Fujii had suggested redoing the
Eötvös experiment. Fujii's work does not seem to have exerted any direct influence
on Fischbach. No citations of it are given in this paper.[21]

The apparatus for the Eötvös experiment is shown in Fig. 1.6 and schematically in
Fig. 1.7.[22] As shown in Fig. 1.7, the gravitational force is not parallel to the fiber due
to the rotation of the Earth. If the gravitational force on one mass differs from that
on the other, the rod will rotate about the fiber axis. Reversing the masses should
give a rotation in the opposite direction. In Fig. 1.6 the horizontal tube typically
contained a standard platinum mass and the vertical tube contained the comparison
mass, suspended by a copper–bronze fiber. Because the experiment was originally
designed to measure vertical gravity gradients, the two masses were suspended at
different heights, which made the apparatus significantly more sensitive to such
gradients than it was to anomalous gravitational accelerations. To calculate and

---

[21]Fischbach keeps detailed chronological notes of papers read and calculations done. He reports
that he has notes on Fujii's work at this time, but does not recall it having any influence on his
work.

[22]Eötvös was originally interested in measuring gravity gradients so the weights were suspended
at different heights. This introduced a source of error into his tests of the equivalence principle, the
equality of gravitational and inertial mass.

Fig. 1.7 A schematic view of
the Eötvös experiment (From
Will 1984)

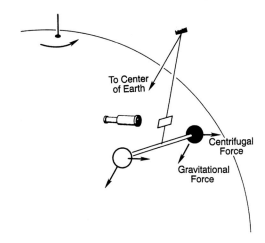

subtract the effect of these gradients Eötvös performed four measurements, with the
torsion bar oriented in the North–South, the South–North, the East–West, and the
West–East directions, respectively. A second apparatus used two torsion balances in
tandem, located on the same mount.[23] In this arrangement the platinum standards
were located at opposite ends of each of the two torsion bars. (For a detailed
discussion of both the apparatus and the methods of data analysis see Fischbach
et al. 1988, pp. 8–12.)

Fischbach attempted to combine the gravitational discrepancy with the energy
dependence of the K^0–$\overline{K}^0$ parameters. They found that if they considered a
hypercharge field with a small, finite mass hyperphoton (the K^0 and $\overline{K}^0$ have
opposite hypercharges) they obtained a potential of the same form as shown above.[24]
They also found that $\Delta k = \Delta a/g$, the fractional change in gravitational acceleration
for two substances, would be proportional to $\Delta(B/\mu)$ for the two substances, where

---

[23] According to one source, the torsion balance was suggested by Juan Hernandez Torsión Herrera
(Lindsay and Ketchum 1962):

> Of Juan Hernandez Torsión Herrera very little is known. He was born of noble parents
> in Andalusia about 1454. He traveled widely and on one of his journeys in Granada with
> his cousin Juan Fernandez Herrera Torsión both were captured by Moorish bandits. Herrera
> Torsión died in captivity but Torsión Herrera managed to escape after a series of magnificent
> exploits of which he spoke quite freely in his later years. During these years he was
> affectionately known as the 'Great Juan' or as the 'Juan Who Got Away.'
> Although not a scientist in his own right, Torsión Herrera passed on to a Jesuit physicist
> the conception of his famous Torsión balance. The idea apparently came to him when he
> observed certain deformations in the machinery involved when another cousin, Juan Herrera
> Fernandez Torsión was being broken on the rack.

> There are some reasons for doubting the veracity of this story.

[24] Fischbach noted that in the limit of infinite range their suggested force agreed with that proposed
earlier by Lee and Yang (1955) on the basis of gauge invariance.

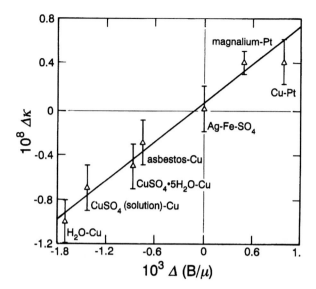

Fig. 1.8 Plot of Δk as a function of $\Delta(B/\mu)$ (From Fischbach et al. 1986)

B was the baryon number of the substance (equal, in this case, to the hypercharge) and μ was the mass of the substance in units of the mass of atomic hydrogen.

They plotted the data reported by Eötvös as a function of $\Delta(B/\mu)$, a quantity unknown to Eötvös, and found the results shown in Fig. 1.8. The linear dependence visible is supported by a least-squares fit to the equation $\Delta k = a\Delta(B/\mu) + b$. They found $a = (5.65 \pm 0.71) \times 10^{-6}$ and $b = (4.83 \pm 6.44) \times 10^{-10}$. They concluded (Fischbach et al. 1986, p. 3): "We find that the Eötvös–Pekar–Fekete data are sensitive to the composition of the materials used, and that their results support the existence of an intermediate-range coupling to baryon number or hypercharge."[25] They calculated the coupling constant for their new interaction for both the Eötvös data and for the geophysical data and found that they disagreed by a factor of 15, which they found "surprisingly good" in view of the simple model of the Earth they had assumed. Not everyone was so sanguine about this, a point we shall return to later.[26]

---

[25]In a later paper (Aronson et al. 1986) the group suggested other experiments, particularly on K meson decay, that might show the existence of such a hyperphoton.

[26]An interesting sidelight to this reanalysis is reported in a footnote to the Fischbach paper. Instead of reporting the observed values of Δk for the different substances directly, Eötvös and his colleagues presented their results relative to platinum as a standard (Fischbach et al. 1986, p. 6): "The effect of this combining say $\Delta k(H_2O–Cu)$ and $\Delta k(Cu–Pt)$ to infer $k(H_2O–Pt)$ is to reduce the magnitude of the observed nonzero effect [for water and platinum] from 5σ to 2σ." $\Delta k(H_2O–Cu) = (-10 \pm 2) \times 10^{-9}$ and $\Delta k(Cu–Pt) = (+4 \pm 2) \times 10^{-9}$, respectively. Adding them to obtain $\Delta k(H_2O–Pt)$ gives $(-6 \pm 3) \times 10^{-9}$.

It seems fair to summarize the Fischbach paper as follows. A reanalysis of the original Eötvös paper presented suggestive evidence for an intermediate-range, composition-dependent force, which was proportional to baryon number or hyper-charge. With a suitable choice of parameters, one could relate this force to anomalies in mine measurements of gravity and to a suggested energy dependence of the parameters of the K^0–$\overline{K}^0$ system.

Fig. 1.9 Plot of Δk as a function of $\Delta(B/\mu)$. *Circles* are data from Fischbach et al. (1986). *Squares* are the final summary from Eötvös et al. (1922). The *dashed line* is the best fit straight line to Fischbach's data. The *solid line* is the fit to the original Eötvös data

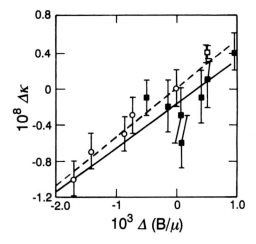

Figure 1.9 shows both the final summary reported by Eötvös as well as Fischbach's reanalysis, along with best-fit straight lines for both sets of data separately (this is my own analysis). Although several of the experimental uncertainties have increased, due to the calculation process, the lines have similar slopes. The major difference is in the uncertainty of the slopes. If one looks at the 95 % confidence level, as shown separately for the Fischbach and Eötvös data, respectively, in Figs. 1.10 and 1.11, one finds that at this level the published, tabulated Eötvös data is, in fact, consistent with no effect, or a horizontal straight line. This is certainly not true for the Fischbach reanalysis.

A skeptic might remark that the effect is seen only when the data are plotted as a function of $\Delta(B/\mu)$, a theoretically suggested parameter. As De Rujula remarked (1986, p. 761): "In that case, Eötvös and collaborators would have carried their secret to their graves: how to gather ponderous evidence for something like baryon number decades before the neutron was discovered." It is true that theory may suggest where one might look for an effect, but it cannot guarantee that the effect will be seen. Although one may be somewhat surprised, along with De Rujula, that data taken for one purpose takes on new significance in the light of later experimental and theoretical work, it is not unheard of.

There is a possibility that Eötvös and his collaborators might actually have seen something of this effect, but discounted it. They report (Eötvös et al. 1922, p. 164): "The probability of a value different from zero for the quantity $x[\Delta k]$ even in these cases is vanishingly little, as a review of the according observational data *shows quite long sequences with uniform departure from the average* [emphasis added], the influence of which on the average could only be annulled by much longer series of observations." The original summary, given in Table 1.1, gives an average value for $x = (-0.002 \pm 0.001) \times 10^{-6}$, which seems to justify Eötvös' original conclusion (Eötvös et al. 1922, p. 164): "We believe we have the right to state that x relating to the Earth's attraction does not reach the value of 0.005×10^{-6} for any of these bodies."

Fig. 1.10 Plot of Δk as a function of $\Delta(B/\mu)$ from Fischbach et al. (1986). The best fit line along with the 95 % confidence level fits are shown

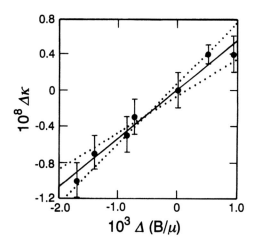

Fig. 1.11 Plot of Δk as a function of $\Delta(B/\mu)$ (The data are from Eötvös et al. (1922)). The best fit straight line along with the 95 % confidence level fits are shown

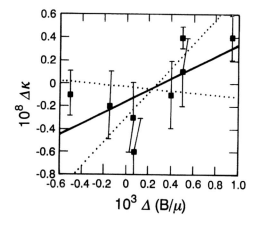

Table 1.1 The value of $x - x_{Pt}$ (Δk relative to platinum) for various substances (From Eötvös et al. 1922, p. 164)

Substance	$(x - x_{Pt}) \times 10^{-6}$
Magnalium	$+0.004 \pm 0.001$
Snakewood	-0.001 ± 0.002
Copper	$+0.004 \pm 0.002$
Water	-0.006 ± 0.003
Crystalline cupric sulfate	-0.001 ± 0.003
Solution of cupric sulfate	-0.003 ± 0.003
Asbestos	$+0.001 \pm 0.003$
Tallow	-0.002 ± 0.003

References

Aronson, S.H., et al.: Determination of the fundamental parameters of the K^0–$\overline{K}^0$ system in the energy range 30–110 GeV. Phys. Rev. Lett. **48**, 1306–1309 (1982)

Aronson, S.H., et al.: Energy dependence of the fundamental parameters of the K^0–$\overline{K}^0$ system I. Experimental analysis. Phys. Rev. D **28**, 476–494 (1983a)

Aronson, S.H., et al.: Energy dependence of the fundamental parameters of the K^0–$\overline{K}^0$ system II. Theoretical formalism. Phys. Rev. D **28**, 495–523 (1983b)

Aronson, S.H., et al.: Experimental signals for hyperphotons. Phys. Rev. Lett. **56**, 1342–1345 (1986)

Bell, J.S., Perring, J.: 2π decay of the K_2^0 meson. Phys. Rev. Lett. **13**, 348–349 (1964)

Bernstein, J., Cabibbo, N., Lee, T.D.: CP invariance and the 2π decay of the K_2^0. Phys. Lett. **12**, 146–148 (1964)

Bock, G.J., et al.: Coherent K_S regeneration on protons from 30 to 130 GeV/c. Phys. Rev. Lett. **42**, 350–353 (1979)

Braginskii, V.B., Panov, V.I.: Verification of the equivalence of inertial and gravitational mass. JETP **34**, 463–466 (1972)

Brans, C., Dicke, R.H.: Mach's principle and a relativistic theory of gravitation. Phys. Rev. **124**, 925–935 (1961)

Carosi, R., et al.: A measurement of the phases of the CP-violating amplitudes in $K_2^0 \rightarrow 2\pi$ decays and a test of CPT invariance. Phys. Lett. B **237**, 303–312 (1990)

Christenson, J.H., et al.: Evidence for the 2π decay of the K_2^0 meson. Phys. Rev. Lett. **13**, 138–140 (1964)

Coupal, D.P., et al.: Measurement of the ratio $(K_L \rightarrow \pi^+\pi^-)/(K_L \rightarrow \pi/\nu)$ for K_L with 65 GeV/c laboratory momentum. Phys. Rev. Lett. **55**, 566–569 (1985)

DeBouard, X., et al.: Two pion decay of K_2^0 at 10 GeV/c. Phys. Lett. **15**, 58–61 (1965)

De Rujula, A.: Are there more than four? Nature **323**, 760–761 (1986)

Eötvös, R., Pekar, D., Fekete, E.: Beitrage zum Gesetze der Proportionalitat von Tragheit und Gravitat. Ann. Phys. (Leipzig) **68**, 11–66 (1922). I have used the translation that appeared in Ann. Univ. Sci. Budap. Rolando Eötvös Nominate, Sect. Geol. **7**, 111–165 (1963)

Fischbach, E.: Tests of general relativity at the quantum level. In: Bergmann, P., De Sabbata, V. (eds.) Cosmology and Gravitation, pp. 359–373. Plenum, New York (1980)

Fischbach, E., Freeman, B.: Testing general relativity at the quantum level. Gen. Relativ. Gravit. **11**, 377–381 (1979)

Fischbach, E., et al.: Interaction of the K^0–$\overline{K}^0$ system with external fields. Phys. Lett. B **116**, 73–76 (1982)

Fischbach, E., et al.: Lorentz noninvariance and the Eötvös experiments. Phys. Rev. D **32**, 154–162 (1985)

Fischbach, E., et al.: Reanalysis of the Eötvös experiment. Phys. Rev. Lett. **56**, 3–6 (1986)

Fischbach, E., et al.: Long-range forces and the Eötvös experiment. Ann. Phys. **182**, 1–89 (1988)

Franklin, A.: The Neglect of Experiment. Cambridge University, Cambridge (1986)

Fujii, Y.: Dilatonal possible non-newtonian gravity. Nature (Phys. Sci.) **234**, 5–7 (1971)

Fujii, Y.: Scale invariance and gravity of hadrons. Ann. Phys. (N.Y.) **69**, 494–521 (1972)

Fujii, Y.: Scalar–tensor theory of gravitation and spontaneous breakdown of scale invariance. Phys. Rev. D **9**, 874–876 (1974)

Galbraith, W., et al.: Two-pion decay of the K_2^0 meson. Phys. Rev. Lett. **14**, 383–386 (1965)

Gibbons G.W., Whiting, B.F.: Newtonian gravity measurements impose constraints on unification theories. Nature **291**, 636–638 (1981)

Grossman, N., et al.: Measurement of the lifetime of K_S^0 mesons in the momentum range 100–350 GeV/c. Phys. Rev. Lett. **59**, 18–21 (1987)

Holding, S.C., Tuck, G.J.: A new mine determination of the Newtonian gravitational constant. Nature **307**, 714–716 (1984)

Holding, S.C., Stacey, F.D., Tuck, G.J.: Gravity in mines—an investigation of Newton's law. Phys. Rev. D **33**, 3487–3494 (1986)

Lee, T.D., Yang, C.N.: Conservation of heavy particles and generalized gauge transformations. Phys. Rev. **98**, 1501 (1955)

Lindsay, D., Ketchum, J.: Unsung Heroes—II: Juan Hernandez Torsión Herrera. J. Irreprod. Res. **10**, 43 (1962)

Long, D.R.: Why do we believe Newtonian gravitation at laboratory dimensions? Phys. Rev. D **9**, 850–852 (1974)

Long, D.R.: Experimental examination of the gravitational inverse square law. Nature **260**, 417–418 (1976)

Long, D.R.: Current Measurements of the Gravitational 'Constant' as a Function of Mass Separation. Nuovo Cimento B **62**, 130–138 (1981)

Mikkelsen, D.R., Newman, M.J.: Constraints on the gravitational constant at large distances. Phys. Rev. D **16**, 919–926 (1977)

O'Hanlon, J.: Intermediate-range gravity: a generally covariant model. Phys. Rev. Lett. **29**, 137–138 (1972)

Panov, V.I., Frontov, V.N.: The cavendish experiment at large distances. JETP **50**, 852–856 (1979)

Rapp, R.H.: Current estimate of mean Earth ellipsoid parameters. Geophys. Res. Lett. **1**, 35–38 (1974)

Rapp, R.H.: Determination of potential coefficients to degree 52 by 5° mean gravity anomalies. Bull. Geodesique **51**, 301–323 (1977)

Roehrig, J., et al.: Coherent regeneration of K_S's by carbon as a test of regge-pole-exchange theory. Phys. Rev. Lett. **38**, 1116–1119 (1977)

Roll, P.G., Krotkov, R., Dicke, R.H.: The equivalence of inertial and passive gravitational mass. Ann. Phys. (N.Y.) **26**, 442–517 (1964)

Scherk, J.: Antigravity: a crazy idea. Phys. Lett. B **88**, 265–267 (1979)

Spero, R., et al.: Tests of the gravitational inverse-square law at laboratory distances. Phys. Rev. Lett. **44**, 1645–1648 (1980)

Stacey, F.D.: Possibility of a geophysical determination of the Newtonian gravitational constant. Geophys. Res. Lett. **5**, 377–378 (1978)

Stacey, F.D., Tuck, G.J.: Geophysical evidence for non-Newtonian gravity. Nature **292**, 230–232 (1981)

Stacey, F.D., et al.: Constraint on the planetary scale value of the Newtonian gravitational constant from the gravity profile within a mine. Phys. Rev. D **23**, 1683–1692 (1981)

Wagoner, R.V.: Scalar–tensor theory and gravitational waves. Phys. Rev. D **1**, 3209–3216 (1970)

Weinberg, S.: Do hyperphotons exist? Phys. Rev. Lett. **13**, 495–497 (1964)

Westfall, R.S.: Never at Rest. Cambridge University, Cambridge (1980)

Will, C.: Theory and Experiment in Gravitational Physics. Cambridge University, Cambridge (1981)

Will, C.: Was Einstein Right? Basic Books, New York (1984)

Zee, A.: Broken-symmetric theory of gravity. Phys. Rev. Lett. **42**, 417–421 (1979)

Zee, A.: Horizon problem and the broken-symmetric theory of gravity. Phys. Rev. Lett. **44**, 703–706 (1980)

Chapter 2
. . . and Fall

2.1 The Immediate Reaction

This suggestion had an immediate impact in the popular press. On January 8, 1986, only two days after the publication of the Fischbach paper, a headline in the *New York Times* announced, "Hints of Fifth Force in Nature Challenge Galileo's Findings."[1] This was the naming of the "Fifth Force."[2] On January 15, an editorial in the *Los Angeles Times* discussed the subject. They cited the skepticism of Richard Feynman, a Nobel Prize winner in physics. Feynman's skepticism concerned the factor of 15 difference (a more careful analysis suggested a factor of 35) between the force needed to explain the Eötvös data and that needed to explain the gravitational mine data. Feynman was bothered more by this discrepancy than Fischbach had been. Feynman expressed this concern in a letter published in the January 23 *Los*

---

[1] This referred to the composition dependence of the suggested force, which implied that different substances would fall at different rates.

[2] The other four forces were the strong, or nuclear, force, the electromagnetic force, the weak force, and the gravitational force.

© Springer International Publishing Switzerland 2016
A. Franklin, E. Fischbach, *The Rise and Fall of the Fifth Force*,
DOI 10.1007/978-3-319-28412-5_2

Angeles Times,[3] and also in a longer and more detailed letter to Fischbach.[4] He agreed that there were possible ways to make the results agree, but regarded them as unlikely. He also questioned the statistical significance of the results claimed in the Eötvös reanalysis. Fischbach answered by pointing out the important effect of local mass asymmetry, discussed below.

The battle would not, however, be conducted or decided either in the popular press or in private correspondence, but rather in the technical literature. During 1986 considerable attention was devoted to the status of the Fifth Force proposal. Questions would be raised as to whether or not the reanalysis was valid and whether or not the hyperphoton idea[5] was already ruled out by K^+ meson decay experiments. Evidence from other previous experiments would also be brought to bear on the subject, and possible theoretical explanations and implications of the new force would be examined. New experiments, as well as improvements in the sensitivity of already existing experiments, would also be suggested.

Another criticism, related to Feynman's, that led to both a refinement of the theoretical model and also to suggested improvements in the sensitivity of Fifth Force experiments was initiated by Thodberg (1986).[6] Thodberg pointed out that

---

[3]Feynman felt that the editorial citation did not convey his real meaning and wrote the following letter. I quote this letter at length not only because it gives a view of the Fifth Force, but also because it illustrates Feynman's views on the methodology of science.

> You reported in an editorial 'The Wonder of It All' about a proposal to explain some small irregularities in an old (1909) experiment (by Eötvös) as being due to a new 'fifth force.' You correctly said I didn't believe it—but brevity didn't give you a chance to tell why. Lest your readers get to think that science is decided simply by opinion of authorities, let me expand here.
>
> If the effects seen in the old Eötvös experiment were due to the 'fifth force' proposed by Prof. Fischbach and his colleagues, with a range of 600 feet it would have to be so strong that it would have had effects in experiments already done. For example, measurements of gravity force in deep mines agree with expectations to about 1 % (whether this remaining deviation indicates a need for modification of Newton's Law of gravitation is a tantalizing question). But the 'fifth force' proposed in the new paper would mean we should have found a deviation of at least 15 %. This calculation is made in the paper by the authors themselves (a more careful analysis gives 30 %). Although the authors are aware of this (as confirmed by a telephone conversation) they call this 'surprisingly good agreement', while it, in fact, shows they cannot be right.
>
> Such new ideas are always fascinating, because physicists wish to find out how Nature works. Any experiment which deviates from expectations according to known laws commands immediate attention because we may find something new.
>
> But it is unfortunate that a paper containing within itself its own disproof should have gotten so much publicity. Probably it is a result of the authors' over-enthusiasm.

This letter was written before the importance of local mass anomalies was pointed out.

[4]Fischbach gave me a copy of this letter.

[5]The hyperphoton was the presumed carrier of the new force.

[6]The editors of *Physical Review Letters* noted a similar letter had also been received from K. Hayashi and T. Shirafuji.

the reanalysis of the Eötvös experiment gave rise to an attractive force, while the geophysical data of Stacey and his collaborators suggested a repulsive force. Fischbach et al. (1986b) stated that Thodberg was indeed correct, but noted that further analysis had shown (p. 2464) "that one cannot in fact deduce from the EPF data whether the force is attractive or repulsive. The reason for this is that in the presence of an intermediate-range force, local *horizontal* [emphasis in original] mass inhomogeneities (e.g., buildings or mountains) can be the dominant source in the Eötvös experiment." In order to determine the magnitude and sign of the anomaly one needed more detailed knowledge of the local mass distribution than was then available.[7] This importance of the local mass distribution could also explain the numerical discrepancy between the force derived from the Eötvös reanalysis and that found from the mine data that had bothered Feynman and others. Similar points were made by De Rujula (1986a), Neufeld (1986), Thieberger (1986), Bizzeti (1986), and Milgrom (1986).[8] These authors suggested redoing the Eötvös experiment by placing the torsion balance on a high cliff, or in a tunnel in such a cliff. They claimed that such a location, which had a large local mass inhomogeneity, could increase the sensitivity of the experiment by a factor of 500.

De Rujula (1986a) and Eckhardt (1986) argued that the original Eötvös reanalysis would not have been sensitive at all to a Fifth Force without local mass inhomogeneities. The argument is that for a deformed rotating Earth the fiber is perpendicular to the deformed surface. For a homogeneous Earth, the symmetry of the local matter distribution will give no net force on the balance. Only if there are local inhomogeneities would the Fifth Force cause an effect. De Rujula quipped (1986a, p. 761): "Although malicious rumor has it that Eötvös himself weighed more than 300 pounds, unspecific hypotheses are not, a priori, particularly appealing."[9] In a note accompanying Eckhardt's paper, Fischbach et al. (1986d) agreed, and pointed out that they had discussed this earlier (see Talmadge et al. (1986) for details).[10]

The initial reanalysis of the Eötvös experiment was partially incorrect because it did not consider local mass anomalies. The subsequent criticism not only modified the theoretical model to stress the importance of these mass inhomogeneities, but also allowed one to design experiments that would be far more sensitive to the presence of the hypothesized Fifth Force.

Other scientists suggested that there was, in fact, no observed effect, that Fischbach had made an error in the reanalysis, and thus, that there was nothing to

---

[7]Detailed calculations based on the local mass distribution were presented in Fischbach et al. (1988).

[8]As discussed later, Fischbach, Talmadge, and Aronson also discussed this at the time and their discussion was later published in Talmadge et al. (1986).

[9]Eötvös was a mountain climber and photographs indicate rather clearly that he did not weigh 300 pounds.

[10]This paper was presented at the March 1986 Moriond Workshop, but did not appear until 1987. The effect of local mass asymmetries was discussed earlier and included in a paper submitted, but not accepted for publication. The collaborators decided to include this calculation in the Moriond paper, even though it was not actually presented at the conference.

Fig. 2.1 Δk as a function of $\Delta(B/\mu)$. These are the results of Keyser et al. (1986) using the raw data of Eötvös and including Renner's data (The *straight line* is from Fischbach et al. 1986a)

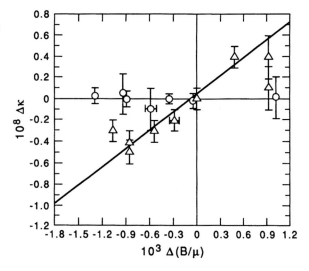

be explained or investigated. Keyser et al. (1986) criticized Fischbach for excluding Eötvös' data on RaBr$_2$, for using the corrected rather than the raw data,[11] and for excluding Renner's (1935) results from their reanalysis of the Eötvös experiment. Figure 2.1 shows their results, which include both Renner's data and the RaBr$_2$ result which used the raw data. At the very least, the apparent effect observed by Fischbach is substantially reduced (compare with Fig. 1.8). They also noted that because the very small RaBr$_2$ sample was in a large brass container, this point was essentially a Cu–Pt measurement, and that it disagreed with the direct Eötvös measurement for Cu–Pt. Fischbach et al. (1986c) responded that there was no essential difference between the raw and corrected data (see Footnote 11), and that reanalysis of both sets of data yielded very similar results. Fischbach also pointed out that Eötvös himself had noted anomalous effects due to heating by the RaBr$_2$ sample, which made that measurement unreliable. In addition, Fischbach found a sign error in the original EPF paper for the RaBr$_2$ point, and noted that after correcting this error, the two measurements agreed, within experimental uncertainty. Fischbach also argued that Roll et al. (1964) had shown that Renner's results were internally inconsistent and that the experimental uncertainties in Renner's results should be multiplied by a factor of 3. This made Renner's work unreliable. In addition, because Renner's

---

[11]In several measurements Eötvös used a brass vial to hold the sample of the material. In reporting the final results Eötvös multiplied the measured value of Δk by a factor $(M_{sample} + M_{container})/M_{sample}$. This assumed that the container had no effect on the measurement. This was a reasonable procedure if one is interested in setting an upper limit, but might overestimate the effect. Similarly, one might also include the composition of the container in calculating $\Delta(B/\mu)$. As Fischbach et al. (1988) later showed, and supported by De Rujula's (1986b) analysis, it makes little difference to the slope of the line Δk against $\Delta(B/\mu)$ whether or not one includes the effect of the brass vials, as long as one is consistent.

experiment was done at a different location, and given the importance of the local mass distribution for the presumed Fifth Force, Fischbach argued that Renner's results should not be plotted on the same graph. I note that this argument applies only if there is an intermediate-range force. If such a force is absent, there would be no reason to exclude reliable data taken at other locations. The point was that Renner's data was unreliable. (See the discussion of this point in the next section.)

Elizalde (1986) agreed with Fischbach that the Eötvös data, when plotted against baryon number or hypercharge, showed a linear relation with a positive slope. He also obtained approximately the same value for that slope. However, he argued (Elizalde 1986, p. 162): "that the errors [on the data points] are so great that it can accommodate the weak equivalence principle of general relativity [no effect] as well as the hypercharge theory of Fischbach et al. (1986a) both for an attractive or a repulsive force." Elizalde cited numerous possible sources of error and uncertainty including impurities in the sample, electric and magnetic fields, local variations in gravity, temperature effects, and others. Although Elizalde presented no details of his calculation, by far the largest effect in the uncertainty in Δk seems to be due to the uncertainty in the abundance of isotopes of the elements (see his Tables 1.1, 2.1 and 2.2). He appears to have made an error. He seems to have incorporated the uncertainty in abundance directly into an uncertainty in $\Delta(B/\mu)$, the quantity of interest. As Fischbach et al. (1988, p. 26) showed later, the uncertainty in (B/μ) due to isotope abundance is approximately 8×10^{-9}, which is negligible compared to the $\Delta(B/\mu) = 0.94 \times 10^{-3}$, for magnalium–platinum.

Kim argued that (1986, p. 255) "the Eötvös experiment cannot provide any conclusive evidence for a new finite range force [...]." He, too, argued that Fischbach had neglected reduced mass effects and had also omitted data, excluded Renner's data, and underestimated the experimental uncertainty. Although Kim was aware of the importance of the local mass distribution for the analysis of the Eötvös result, he did not cite Fischbach et al.'s (1986b) discussion, and still argued that the difference in sign between the Eötvös and geophysical results was a serious problem. Kim also noted, as discussed below, that others had argued that K^+ meson decay experiments seemed to rule out the hyperphoton explanation. He concluded that the variations in Δk were consistent with a null result within one or two standard deviations, but did admit that there did seem to be a linear relationship between Δk and $\Delta(B/\mu)$.[12] Both Elizalde and Kim cited an Eötvös result on magnalium and platinum that gave $\Delta k = 0.6 \times 10^{-8}$ relative to the Sun, and 0.4×10^{-8} relative to the Earth, as arguing against a finite range force, which would affect the Earth measurement but not that relative to the Sun. It is true that the difference is consistent with zero, but it is also consistent with a positive result for a finite-range force.

De Rujula (1986a,b) remarked on the possible effect of reduced mass (see Footnote 11), and took it into account in his own reanalysis. He also found a linear

---

[12]He discounted this effect because he assumed that there were different systematic errors for different parts of the Eötvös data. He did not explain how these would give rise to a linear effect, nor did he consider whether or not these errors would affect the null result he favored.

relation between Δk and $\Delta(B/\mu)$ and found $\Delta k = (5.63 \pm 0.68) \times 10^{-6}\Delta(B/\mu)$. This is quite close to the Fischbach result[13] $\Delta k = (5.65 \pm 0.71) \times 10^{-6}\Delta(B/\mu) + (4.83\pm6.44)\times10^{-10}$. The agreement between De Rujula's and Fischbach's analyses confirms Fischbach's calculation that showed that the value of the slope of the line (Fischbach et al. 1988, p. 29) "is nearly the same irrespective of whether one uses the data for the 'composite' values of the samples [...] (brass vials included) or the 'reduced' values of the samples [...] (vials not included)."[14]

Aronson et al. (1986) suggested that one might look for the hyperphoton, the carrier of the force, by searching for the decay K^+ meson $\rightarrow \pi^+ + \gamma_Y$, where γ_Y is the suggested hyperphoton. They noted that existing limits on this decay set severe constraints on both the range and strength of the proposed force. Other physicists had different views on this issue. Suzuki (1986), Bouchiat and Iliopoulos (1986), and Lusignoli and Pugliese (1986) examined the same decay and concluded that the experimental limits were far lower, by a factor of approximately 1000, than the theoretical predictions of the Fifth Force. There were theoretical assumptions in these calculations. As Bouchiat and Iliopoulos remarked (p. 449): "We conclude that, unless one is willing to abandon all our ideas on PCAC and assume totally unrealistic values for the extrapolation factor x, the model is already in contradiction with the data by several orders of magnitude."[15]

Fig. 2.2 Plot of Δk as a function of $\Delta(Z/\mu)$ from De Rujula (1986b). No apparent effect is seen

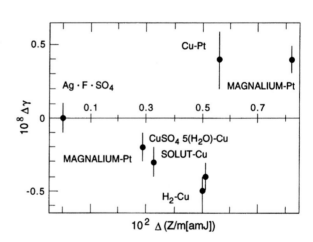

---

[13]The analyses done by De Rujula and Fischbach were slightly different. Fischbach included a constant term in his fit, which, as one can see, is very small.

[14]De Rujula also plotted Δk against $\Delta(Z/\mu)$, where Z is the atomic or lepton number. As shown in Fig. 2.2, he found no effect. He seems to have regarded this as casting doubt on the Fischbach reanalysis.

[15]This is an example of the Duhem–Quine problem. As Quine pointed out, any statement can be held to be true if one is willing to make modifications elsewhere in one's knowledge. Bouchiat is stating that he thinks some modifications are unreasonable. One has to give up too much.

Neufeld (1986) also pointed out than an earlier experiment by Kreuzer (1968), which had been interpreted as showing that the ratios of "active" to "passive" gravitational mass of two substances differed by less than 1 part in 20,000, could also be used to set an upper limit on an intermediate-range force associated with baryon number, i.e., the Fifth Force.[16] He found that if he interpreted Kreuzer's "upper bound" as a one standard deviation limit, then it would be consistent with the Eötvös reanalysis within two standard deviations. Nussinov (1986) agreed that Kreuzer's experiment was consistent with the Fischbach reanalysis and suggested that more precise versions of the experiment be done. He remarked that more precise satellite versions of the Eötvös experiment were being planned.

Bertolami (1986) obtained a limit on the range of the Fifth Force from a reanalysis of the quantum gravity experiment of Colella et al. (1975). For a 1 % fringe shift and using $\alpha < 10^{-2}$ from geophysical measurements, he found $\lambda < 60$ m.[17]

Another question was raised as to whether or not one could explain the Eötvös reanalysis in terms of more conventional physics, without invoking a new force. In the original Fischbach et al. paper (1986a), it had been noted that Dicke had asked whether a systematic effect dependent on thermal gradients (different temperatures at different parts of the apparatus) and the densities of the substances could explain the observed correlation between Δk and $\Delta(B/\mu)$. Fischbach responded that their investigation indicated that it couldn't.[18] Chu and Dicke (1986) presented a more detailed model (p. 1823): "We find that systematic effects due to thermal gradients can account for the experimental data." Fischbach et al. (1986d) found the idea of a thermal effect fixed in both magnitude and direction, and constant over a period of years, the duration of the Eötvös measurements, very unlikely. They also pointed out that the Chu–Dicke model had difficulty in explaining the data taken with platinum. They remarked somewhat later, however, that (Fischbach et al. 1986f, p. 1113) "the C–D [Chu–Dicke] model is very clever and is sufficiently promising to warrant more detailed study."

Although the criticism may have made the Fischbach reanalysis somewhat uncertain, it did not prevent physicists from planning new, more sensitive versions of old experiments and designing new ones to test the presence of the Fifth Force.[19] As discussed earlier, Thieberger, Bizzeti, Neufeld, and Milgrom all suggested improving the sensitivity of the Eötvös experiment to the presence of the Fifth Force by locating the apparatus at a site near a large mass inhomogeneity such as a hillside or a cliff. Similarly Speake and Quinn (1986) suggested using a beam

---

[16]"Active" gravitational mass is the mass that is the source of a gravitational field, while "passive" mass is the mass on which the gravitational force acts.

[17]Fischbach points out that this experiment was not, in fact, sensitive enough to set such a limit.

[18]Details appeared in Fischbach et al. (1988).

[19]I am discussing here the context of pursuit which involves using the hypothesis as the basis for further investigation. As discussed below, scientists who investigated the Fifth Force had varying attitudes toward its truth.

balance with weights made of two different substances at two different locations, one very asymmetric, as a test of the hypothesis.

Another suggestion to test the composition-dependence of the Fifth Force was to repeat Galileo's mythical Leaning Tower of Pisa experiment by dropping masses of two different substances and observing whether or not they fell at the same rate (Cavasinni et al. 1986). Others suggested making use of new space technology to observe possible effects. These included observing possible effects on two small balls orbiting each other in the Space Shuttle (Avron and Livio 1986), or the effect on larger masses such as orbiting spacecraft (Schastok et al. 1986; Hills 1986). A review of other suggested Fifth Force experiments as well as tests of the inverse square law appeared in Paik (1986).

Theorists were also busy. Earlier calculations had argued that the proposed hyperphoton mechanism was incompatible with the known K^+ meson decay data. Other theorists attempted to provide an explanation of the Fifth Force based on fundamental interactions. Fujii (1986) examined various models that might provide an explanation including his own previous suggestion of a scalar–tensor theory. Pimental and Obregon (1986) used a similar approach and found a result that in the weak-field limit had the same mathematical form as that suggested by Fischbach. Other approaches were also tried. Fayet (1986) used supersymmetry[20] to derive an intermediate-range force, while Hayashi and Shirafuji (1986) obtained a similar result using fermion coupling. A similar result was found by both Bars and Visser (1986) and by Barr and Mohapatra (1986) from a "compactification" of a five-dimensional Kaluza–Klein theory. Bars and Visser also noted that observation of a Fifth Force effect might provide evidence for higher dimensions. Moffat (1986) showed that deviations from Newtonian gravity were compatible with his nonsymmetric theory of gravitation.

Some theorists looked at the implications of the Fifth Force in other areas. These included stellar energy loss (Grifols and Masso 1986), hyperphoton production in intermediate vector boson decay (Rizzo 1986), neutron–antineutron conversion (Massa 1986), and radiation of particles from binaries (Li and Ruffini 1986). In all cases the effects were quite small and did not suggest new experimental tests.

At the end of 1986 the evidential context for the Fifth Force was much the same as it had been at the beginning of 1986 when Fischbach et al. had first proposed it. By early 1986 the inverse square law of gravitation had been tested at very short distances and was confirmed. This disagreed with Long's earlier work. Newman and collaborators (Hoskins et al. 1985) had compared the measured ratio of torques to that expected from Newtonian theory at distances of 5 and 105 cm. They found $(R_{\text{expt}}/R_{\text{Newton}} - 1) = (1.2 \pm 7) \times 10^{-4}$. Assuming that the violation had the same

---

[20]Ordinary particles are divided into fermions, with half-integral spin, 1/2, 3/2, etc., and bosons with integral spin. Supersymmetry suggests that each particle has a supersymmetric analogue in the other class. Thus, an electron will have a supersymmetric analogue, the selectron, which instead of having spin 1/2 will have integral spin. Similarly, the spin one photon will have an analogue, the spin 1/2 photino.

form as Long's $(1 + \epsilon \ln r)$, this gave $\epsilon = (0.5 \pm 2.7) \times 10^{-4}$. They also presented a more detailed discussion of their earlier experiment (Spero et al. 1980) that had found $\epsilon = (1 \pm 7) \times 10^{-5}$. Holding et al. (1986) had also published a paper giving details of their mine measurements that had provided the values of α and λ cited informally in Fischbach's original paper. Strong doubts had also been raised about the proposed hyperphoton mechanism, but other explanations were possible, as discussed earlier.

The attitude of scientists toward the Fifth Force at this time varied from outright rejection to considering it highly suggestive and plausible. Glashow, a particle theorist, was quite negative (quoted in Schwarzschild 1986, p. 20): "Unconvincing and unconfirmed kaon data, a reanalysis of the Eötvös experiment depending on the contents of the Baron's wine cellar [an allusion to the importance of local mass inhomogeneities], and a two-standard-deviation geophysical anomaly! Fischbach and his friends offer a silk purse made out of three sows' ears, and I'll not buy it."

Others cited the fruitfulness of the suggestion. Bars and Visser cited Fischbach's analysis as the original motivation for their work, but noted that (Bars and Visser 1986, p. 25): "It now appears that there are potentially serious problems with the analysis as advanced by Fischbach et al." They were referring to the difficulties discussed earlier about the sign and magnitude of the proposed Fifth Force (1986, p. 25): "Nevertheless, in the course of our work we became convinced that forces similar to the reported one are likely to exist as remnants of higher dimensions." Even if wrong, they regarded Fischbach's suggestion as having had a positive effect. Maddox noted that (1986a, p. 173), "Fischbach et al. (1986a) have provided an incentive for the design of better measurements by showing what kind of irregularity it will be sensible to look for." An important feature of experimental design is knowing how large the observed effect is supposed to be.[21]

A much more positive view was (Lusignoli and Pugliese 1986, p. 468): "Considerable, and justified, excitement has been provoked by the recent announcement [by Fischbach]—that a reanalysis of the celebrated Eötvös experiment together with recent geophysical gravitation measurements supports the existence of a new fundamental interaction."

Paik's summary of the situation seems reasonable (Paik 1986, p. 394):

> It is clear that the recent announcement of the possible discovery of a 'Fifth Force' (Fischbach et al.) stimulated great interest on the part of experimentalists to resume, improve, and accelerate old experiments, as well as to plan new experiments. After the storm of criticisms, the essential claim of Fischbach et al. that the original Eötvös data show a strong correlation with chemical composition seems to be intact. Whether this represents a new physics or is an artifact of statistical fluctuation, only time will tell.

It seems clear, judging by the substantial amount of work published in 1986, that a significant segment of the physics community thought the Fifth Force hypothesis was plausible enough to be worth further investigation. This was about to become even more apparent. Although almost invisible in the published literature,

---

[21] I will discuss later the idea of an enabling theory, one that assists in experimental design.

experiments were being designed, performed, and analyzed. The results would start
to appear in early 1987.

2.2   An Electronic Interlude

In the previous section I discussed the immediate reaction of the physics community
to the suggestion of a Fifth Force in gravity along with the responses of Aronson,
Fischbach, and Talmadge to the comments and criticism, as they both appeared in
the published literature.[22] One of the possible dangers of such an approach is that it
gives too "sanitized" a history. One might worry that the public arguments were not
those that the scientists used privately.

In this section I will present evidence to support the view that this is not a
significant problem. I will examine the electronic correspondence between Aronson
and Fischbach and Talmadge for the period 16 January 1986 to 22 July 1986. The
original Fifth Force paper was published on 6 January 1986 so this correspondence
covers the period immediately following the publication of the idea. During this
period Aronson, Fischbach, and Talmadge continued to work on the Fifth Force,
ultimately producing a long, detailed account of their work (Fischbach et al. 1988).
I will examine both this work as well as their reactions and responses to several
of the comments and criticisms, discussed above, that were offered at this time.
That the electronic mail is not a complete record is indicated by references to both
letters and telephone calls. It is, nevertheless, a fascinating and honest look into
the practice of science. None of the participants had any idea of the eventual use
of this correspondence. In addition, because this electronic mail contained all of
Aronson's electronic correspondence, which included far more than just the Fifth
Force communications, it gives an insight into the professional life of a practicing
senior scientist. I will begin, however, with the Fifth Force.

The first issue raised concerned the related problems of the relative sign of the
Eötvös effect and the geophysical anomaly (were the forces needed attractive or
repulsive?) and the question of the effect of the local mass distribution on the Eötvös
effect.[23] The episode began when De Rujula, a theorist at CERN, asked, during a
seminar being given by Aronson, whether or not the sign of the Eötvös effect agreed
with the repulsive force required by the geophysical data (Aronson, 20 Jan). The
next day Aronson reported that he had received a telephone call from Thodberg who
noted that the reanalysis of the Eötvös experiment seemed to require an attractive
force, while the geophysical data indicated a repulsive force. Aronson queried his

---

[22]I am grateful to Sam Aronson for providing me with a hard copy of the electronic mail used in
this section. I should emphasize that neither he, nor Ephraim Fischbach or Carrick Talmadge, has
made any suggestion as to how I might use this material.

[23]In what follows I will cite the electronic mail correspondence by giving the author of the letter
and the date.

collaborators (21 Jan): "If all this is correct and our Eq[uation] (3.4) doesn't have a sign error, haven't we found an attractive force?" Thus, there was a possibility that the whole enterprise rested on an error. Much of the support for, and plausibility of, a Fifth Force had come from the fact that the *same* force could explain the Eötvös data, the geophysical anomaly, and the $K^0 - \overline{K}^0$ energy dependence. A written version of Thodberg's comments followed and Aronson replied (22 Jan): "I agree that Eq. (3.4) is confusing; it basically only gives the magnitude of the ratio delta $(a)/g$. The question of the sign of the effect, while all-important, cannot be unambiguously inferred from the Eötvös paper." Aronson further reported (24 Jan.) that he had received a copy of Thodberg's comment, sent to *Physical Review Letters (PRL)*, which argued that one didn't have the freedom to change the sign of the Eötvös anomaly to agree with the geophysical data. Fischbach responded to both Aronson and, in a telephone call, to Thodberg (25 Jan): "In principle, the Eötvös experiment *can* [emphasis added] unambiguously fix the overall sign, however its unclear whether Eötvös actually worried about the sign in their write up. We have fixed the sign using the geophysical data [. . .]." It seems clear that Fischbach was unable to convince Thodberg because Thodberg's comment was subsequently published in *PRL*.

Thus, the question of what Eötvös and his collaborators had actually done and said had become extremely important. Aronson asked Talmadge for a copy of an English translation of the Eötvös paper (the original was in German) that was then being done at the University of Washington.

The debate about the sign of the effect became academic when Talmadge reported that the effect of an asymmetric local mass distribution, assuming there was a nearby mountain, would swamp the usual effect.[24] It was both larger, by a factor of 10,000, and could be of either sign depending on the local mass distribution. He noted (28 Jan): "It does appear that our statement [contained in the original Fifth Force paper, Fischbach et al. (1986a), which mentioned the importance of the local mass distribution] may be true with a vengeance." Aronson confirmed Talmadge's calculation and suggested including the result in their response to Thodberg, which, as discussed above, was done.

On a more personal level, Aronson worried that his collaborators might have thought that he had lost faith in their joint work (Aronson, 29 Jan):

> I hope you're taking my reaction to the sign controversy in the right spirit; sometimes I must seem to have gone off the reservation! Don't forget I'm sitting here in a very large concentration of aggressive and high-powered physicists and getting a fair amount of close questioning from them on a daily basis. Probably you're getting it too but perhaps at more of a distance. In any case I want very much to preserve the credibility of our work and being remote from you [Fischbach] and Carrick [Talmadge] (who have done most of the work) I wind up simply passing some of the critical attitudes on to you. I'm sorry for that but getting people here working on the problem is in my view worth the extra trouble I'm causing.

---

[24]It was still being assumed, incorrectly, that there would be an effect of a Fifth Force in the Eötvös experiment for a symmetrical Earth.

Work continued on the problem in order to complete the more detailed account. Aronson reported that he had spoken with Judith Németh, a Hungarian physicist in Budapest, concerning the actual local mass distribution—buildings, gardens, basements, etc.—around the laboratory where Eötvös did the experiment. Unfortunately (6 Feb), "As she says about the campus geography, whatever we can get may not be entirely correct; the buildings are not interesting enough from a historical or architectural point of view to have been well-documented." Németh later sent Aronson a hand drawn map of the campus giving the locations of various buildings.

On 9 February, Aronson received the text of a paper on local mass asymmetries that the collaboration proposed to send to *PRL*. This paper dealt with the effects of local mass asymmetries and presented an approximate calculation of the size of the effects for a model of the location of the Eötvös experiment. They assumed a four-story laboratory building, with each floor 30–50 m $\times$ 30–50 m $\times$ 3 m, and similarly for a basement. They found that the effect of the building was comparable to that expected for a uniform, spherical Earth, and that the effect of the basement, or hole in the uniform mass distribution, was several times greater. They concluded (Talmadge, 9 Feb): "It follows [...] that for the actual conditions of the EPF experiment, modeled as we have here, the sign of the EPF results would correspond to a 'repulsive' force [...]." The numerical value of the result also eliminated any serious discrepancy between the Eötvös and geophysical results.[25] The problem had been solved, at least for the collaborators.

The fact that a solution was available did not stop work on the problem. On 18 February, Aronson reported that he had again spoken with Judith Németh. A question had been raised earlier about some experiments Eötvös had done at Lake Balaton. Németh reported that they had nothing to do with the 1922 publication.[26] She also informed Aronson that there was a cellar under the entire building in which Eötvös had performed his experiments, an important point for estimating the effect of local mass asymmetries. Talmadge further examined topographical maps of Budapest and found a hill on the other side of the Danube from Eötvös' laboratory that might give a significant effect if the range of the force were larger than 200 m (26 March). Other scientists also became involved. Aronson reported that Kiraly, a Hungarian physicist, was trying to estimate the effects of different locations within the building Eötvös had used.

Not all good deeds are immediately rewarded. On 30 April, Talmadge reported that their mass anomalies paper had been rejected by *Physical Review Letters*. It seemed that four papers on the subject had been submitted and that *PRL* had decided to publish only the first, that of Thieberger (1986). Two others, Milgrom (1986) and Bizzeti (1986), were published elsewhere. The work done by Talmadge et al. was later included in their detailed account, Fischbach et al. (1988). Talmadge also

---

[25]The authors also noted that for a deformed, rather than a spherical, Earth, the effect on the Eötvös experiment would be rigorously zero.

[26]Aronson also remarked that (18 Feb) "it might be interesting to find those [the Lake Balaton experiments] but let's leave that to future generations of paleophysicists."

reported that a first draft of this paper was nearing completion. The calculation also appeared in the written version of the talk that Aronson had presented at the March 1986 Moriond meeting (Talmadge et al. 1986). Aronson had not, in fact, discussed this calculation at the conference, but the collaborators decided, with the approval of the conference organizers, to include it in the written version. A major reason for this was to include a calculation that contained a detailed model of local mass asymmetries, those in the vicinity of Eötvös' laboratory, something that was not presented in any other papers.

Fischbach et al. (1986f) summarized the situation as of May 1986, before any new experimental results were available, in a talk presented at the Lake Louise conference on particle and nuclear physics. He concluded (p. 1105): "Despite the uncertainties in the geophysical and K^0–$\overline{K}^0$ data, and other uncertainties in the EPF data [. . .], the observation that effects appear in all three systems, which are at least roughly compatible with the same potential [. . .], must be taken seriously." His talk also summarized the status of the criticisms and comments and stated (p. 1108): "from what we know at present there are no outstanding criticisms which challenge the basic assertions contained in Ref. 1 [the original Fifth Force paper, Fischbach et al. 1986a]." Not everyone would have agreed with Fischbach's optimistic evaluation at that time, as witnessed by the continued publication of criticisms discussed earlier. A note of caution was also present (p. 1105): "Nevertheless it could very well develop in the end that all of these effects are spurious and that $U_Y(r)$ [the Fifth Force potential] does not exist."

The communication within the collaboration concerning the written version of this paper was intense. Aronson received no fewer than seven versions of the paper within a two week period, on 22–26 June and on 1, 2 July. There was considerable discussion of the wording of the paper and how the conclusions were to be presented. In particular, the criticism by Keyser et al. (1986) concerning the exclusion of Renner's data was debated. Version I (22 June) read (Talmadge, 22 June):

> 3) Keyser, Niebauer, and Faller (KNF) raised the question of why we excluded the data of Renner, who repeated the experiment with a modified version of the original apparatus in 1935. The reason for this, as we noted in Ref. 1, was that Roll, Krotkov, and Dicke (RKD) found various inconsistencies in Renner's analysis and results. Renner's data are discussed in much greater detail in our longer paper, but since KNF in no way refute the arguments of RKD, one must be very careful in using these data. Moreover, even if Renner's data were free of problems, it would still be incorrect to directly compare them to the EPF data since the two experiments were carried out in different places. Ref Barnothy [a private communication from a Hungarian physicist]. As we noted previously, the measured values of 'Delta kappa' depend sensitively on the local matter distribution. Hence it is entirely conceivable that Renner could have seen an (almost) null result, which is what the data would suggest if they were correct, despite the fact that EPF found a nonzero effect.

Aronson did not approve (Aronson, 23 June):

> 2) There is a statement about Renner's data which I also didn't like in the long paper. You say something to the effect that, having done the experiment somewhere else, Renner could have gotten a different result. In particular he could have gotten a null result, which his data (were they any good) imply. I think this is pushy; it's eating your cake and having it, even

if it is strictly correct. It gives the impression that after taking RKD seriously, we would be ready to accept Renner's data if we had reason to believe he had gotten a null result. I think this isn't so; I think Renner's data are [. . .] forever useless.

Despite, or perhaps because of, modern technology, Version II reached Aronson before any action on his comments could be made. He noted that (24 June) "the comments I made on Version I still apply to Version II." In a letter sent along with Version II, Talmadge and Fischbach asked Aronson to suggest a rewording of the material. Talmadge wrote (23 June):

Regarding Renner, Ephraim [Fischbach] and I both feel that it is important that we get across the message that it is perfectly consistent with the hypothesis of a hypercharge field that Renner found neither the same magnitude nor sign for his slope parameter (at the least, using our current model for the local environment, the different sign is certainly consistent with what one would expect). This isn't to say that the language can't be toned down, especially if you think it is misleading as is. Ephraim suggests that if you have time, maybe you could try reworking that part of the paper to improve the language [. . .].

Aronson proposed the following modification, to appear after the reference to Barnothy (see above) (24 June):

In the presence of nearby sources whose strengths are comparable to that of the Earth, two experiments (such as EPF and Renner) at different locations could obtain correlations of delta kappa to delta(B/μ) which differ in magnitude and even in sign. In the case of these two experiments, the differences in local mass distributions are not well-known. This would prevent our combining or comparing their results even if serious problems with Renner's analysis did not exist.

He elaborated on the reasons for his desired changes (24 June):

What, you ask, is the big deal? Maybe I'm too touchy, but I felt the reference to Renner's null result as a possible result, while at the same time implying that his analysis is fubar, puts us in a position where we seem ready to accept any result that fits our hypothesis. I guess we should convince ourselves we wouldn't have combined EPF's and Renner's results even if they WERE statistically compatible. [I note here Aronson's earlier comment, given above, that Renner's data were forever useless]. In any case I feel more comfortable simply claiming that under different conditions different results can come out of such experiments. However, I'm not married to the verbiage presented above; it's the thought that counts.

Talmadge and Fischbach both liked the new wording better than the original and Version III and all subsequent versions, including the published text (see Fischbach et al. 1986f, p. 1110), included the modification.[27]

This discussion among the collaborators concerned not only the best way to present their argument, but was also about what constituted good data and the evidential weight of that data. Thus, it was important to emphasize that the differences in local mass asymmetries at different locations could give rise to different results, even to null results. The unreliability, and hence the small evidential weight, of Renner's data also needed emphasis. These were related issues because of the

---

[27]A more detailed analysis of Renner's data appeared later in the long paper (Fischbach et al. 1988).

different location of Renner's experiment, but they were not identical. It was the separation of the two issues and their clarification that formed the discussion.

The question of whether or not to include data, whether or not it is "good" data, is one of the most important issues in performing an experiment and reporting its results. Almost all experimenters will exclude some data because the experimental apparatus was not operating properly, there were background problems, etc. This is even more difficult when the experiment was done by someone else 50 years previously. Thus, there was a significant amount of effort devoted to evaluating Renner's data. (For a detailed discussion of a similar issue, that of Millikan's exclusion of data in his oil drop experiment, see Franklin 1986, Chap. 5.)

The importance of data evaluation and selection also appeared in discussions of the "talg", "schlangenholz", and $RaBr_2$ points in the reanalysis of the original Eötvös data. The discussions of "talg" (tallow) and "schlangenholz" (snakewood) concerned the chemical compositions of the two substances, necessary quantities for evaluating (B/μ). Talmadge wrote to Aronson (23 June): "The 'talg' point is less certain—it could mean tallow, suet, grease [. . .]. If we take the point to be 'tallow' (which is reduced beef fat), then since tallow is basically fatty acids+glycerol and water then [. . .]" the value of (B/μ) would be between 1.00680 (fatty acids) and 1.00731 (glycerol). (Water is 1.00723.)[28] The uncertainty in B/μ made it unclear how this datum should be used. Talmadge wrote (25 June): "Regarding the tallow datum, our [Fischbach and Talmadge] position is that we really don't feel that the composition of tallow can be established with enough certainty to put it on an equal weighting with the other data points. Hence the straight line represents a fit which excludes this datum [. . .]."

Aronson responded (26 June):

> I'm afraid I disagree with you all on the tallow point. There is nothing to appear to be hiding about this datum. We can do the fits with the tallow point twice, at each end of the reasonable range of B/μ for that point. I bet there is very little effect on the fit anyway. One can give the impression of over-managing the data, and for what gain? It's like the question of the $RaBr_2$ point; we have to decide if it is good data and then include it or not. In that case I think we agree it is subject to two or more systematic effects that we can't estimate. Here we have some small uncertainty as to where to place the point in the context of our model; we have no reason to doubt the measurement of delta kappa itself. I say include it in the fits.

Talmadge and Fischbach agreed (26 June):

> Regarding the tallow datum, I agree with what you've said completely. The only problem is that neither Ephraim nor I have the slightest idea how to set a realistic upper bound on the value of (B/μ) for tallow. If EPF meant the kind of tallow that one cooks with, it is probably pretty easy, since chances are that the stuff is pure; but if (as the German post docs here felt) it is the stuff candles were made of, who knows what in the world was added to the beef or mutton fat to produce the final material that EPF called "talg?" In any case, I should point out that it makes our argument easier, if we were to include the tallow datum,

---

[28]The group ultimately decided on a value of $(B/\mu) = 1.00691$. See Fischbach et al. (1988, p. 38) for details.

since (3.1) the quality of the fit is reduced to a more realistic level, and (3.2) the magnitude of the slope is decreased, making it easier for our present model of the mass distribution to explain EPF's results in terms of the nominal values of alpha and lambda. A similar thing happens if you include the RaBr$_2$ datum, of course.

Aronson attempted to conclude the discussion (27 June):

> One last shot on included and unincluded data; don't let yourself be swayed by whether including a particular point helps or hurts our case. We have to decide only if it's good data or not. I believe the RaBr$_2$ point is not a reliable measurement of delta kappa and that the talg point is. We then have to figure out how to include it; how far does the point move in delta B/μ if talg is candle wax (say paraffin)? I think we're better off including it and discussing the (negligible) effect of that point's position on the result, than excluding it and inviting infinite Talmudic debate on how many red herrings can dance on the end of a candle.

There were similar discussions concerning the chemical composition of "schlangenholz" (snakewood). Ultimately, chemical analyses were performed on two different samples, one from Brazil and one from Surinam, which both gave $(B/\mu) = 1.00750$, within experimental uncertainty. The RaBr$_2$ data was excluded because it was not considered a reliable result (see earlier discussion).

The graph presented in the Lake Louise paper included the snakewood–platinum point as a good data point, while the tallow–Cu point was shown with a different symbol (Fischbach et al. 1986f, p. 1107) "to indicate an uncertainty in the value of (B/μ) for tallow." In the complete analysis presented in Fischbach et al. (1988) three different fits to the data are given: (3.1) using all the data points save RaBr$_2$, snakewood, and tallow, (3.2) using all the data from (3.1) plus snakewood, and (3.3) using all the data from (3.2) plus tallow. The results, shown in Table 2.1, are not significantly different, nor do the confidence levels in the fit change significantly. They regarded the asterisked line as their best values. This included the snakewood measurement, but not tallow.

The email record also included discussions of an alternative explanation of the Eötvös results, the thermal gradient proposal of Chu and Dicke (1986). On 11

Table 2.1 Results of various fits between the recalculated values of Δk and $\Delta(B/\mu)$ (From Fischbach et al. 1988). $\Delta k = \gamma \Delta(B/\mu) + \delta$

Points fitted	$10^6 \gamma$	$10^9 \delta$	x^2/(d.o.f.)	% C.L.
Reduced values				
(1)	4.82 ± 0.62	0.16 ± 0.61	1.2/5	94
(2)*[a]	4.81 ± 0.62	0.30 ± 0.59	2.0/6	92
(3)	4.56 ± 0.57	0.32 ± 0.59	3.1/7	88
Composite values				
(1)	4.81 ± 0.69	0.06 ± 0.44	1.3/5	94
(2)	4.78 ± 0.69	0.12 ± 0.44	2.2/6	90
(3)	4.49 ± 0.64	0.13 ± 0.44	3.5/7	83

[a]They regarded the asterisked line as their best values. This included the snakewood measurement, but not tallow

March, Talmadge wrote that he had just received a letter from Dicke claiming that the Eötvös results could be explained by thermal effects.[29] Talmadge did not think that the fit to the data (χ^2 of 13 for 7 degrees of freedom, a probability of approximately 7 %)[30] was particularly good, nor did he understand how a thermal effect could mimic the (B/μ) correlation. More details followed (Talmadge, 12 March). Dicke had fitted a selected sample of the data looking for a correlation with the inverse of the density of the sample, or with its area, and obtained what he regarded as acceptable fits.[31]

Fischbach and Talmadge offered several questions and objections (12 March): "[The] main objection seems to be that we don't understand how it is possible to construct a realistic model in which the thermal gradients couple as strongly as Dicke's model would suggest." They also noted that it was no surprise that Dicke's model gave a reasonable fit to the data points that used copper as a comparison standard because (B/μ) increased smoothly as a function of atomic number, and any other quantity which had the same property, such as (electron number/μ) or inverse density, would also give a good fit, provided that the data did not span the double-valued nature of the B/μ curve, as Dicke's selected data points did not. Talmadge also reported that it was rumored that it was a common practice at the time of the Eötvös experiment to multiply one's error by a factor of the square root of pi. If this were, in fact, the case then Dicke's model would give a χ^2 of 38.8 for 7 degrees of freedom (which had a probability of less than 0.07 %) and would thus be ruled out on statistical grounds.[32] They also raised a question concerning Dicke's data selection, which seemed to them to be too selective.

Work continued on this problem. Talmadge wrote to Aronson that he had been unable to reproduce either Dicke's results or his χ^2. He noted that, because of the shape of the (B/μ) curve (19 March), "the only data points which actually put a real test to Dicke's model are the data points with the platinum standard, and it is *precisely* these points that his model appears unable to explain." Aronson further queried Dicke's explanation (24 March):

> Have we ever understood how, physically, a thermal gradient effect which depends on the length of the samples actually produces a net torque on the balance? [...] I think you need a physical hypothesis to test, not just a formula with enough parameters to fit the data. It is

---

[29] Recall that the original Fifth Force paper (Fischbach et al. 1986a) had stated that Dicke had raised this question.

[30] This was the fit contained in the letter. The published values were different depending on the data used and the method of fitting.

[31] The Eötvös measurements were made using three different methods. Method I assumed that both the torsion constant and the gravity gradients remained constant during the observations. Method II allowed the torsion constant to very slowly, and Method III allowed both the torsion constant and the gravity gradients to change with time. The measurements also used either a single-arm or double-arm torsion balance. (See Fischbach et al. (1988, pp. 11–12) for details.) Dicke thought that the different methods had different systematic effects and fitted the single-arm and double-arm data separately.

[32] Talmadge also noted that this might explain the exceptionally good fits that they had obtained.

true that Dicke doesn't have as many parameters as data, so there is statistical significance to his fit, but why do we ascribe it to thermal effects?

Talmadge and Fischbach agreed that there didn't seem to be a physical basis for Dicke's model and offered several possible explanations of their own, none of which seemed to work. Aronson further noted that he *had* been able to reproduce Dicke's fit to the data (26 March):

> On Carrick's claim that he couldn't reproduce Dicke's χ^2 even by scaling from Dicke's graph; I did the same and get reasonable good agreement—about $\chi^2 = 12$ or 13 for each of the two figures. I don't know if this is a big issue; more important may be that Carrick gets different values for a similar looking fit. Still, as I said in my last note, the most important point also pointed out by Carrick is that to fit the data Dicke has to come up with correlations of opposite signs to thermal effects for the different EPF setups.

Talmadge responded that he had indeed found an error in his own calculations. He had included a single data point twice. He added that Dicke had also made a small error in using an incorrect value for the experimental error on one point. All the fits now seemed to agree. He also reported that a member of Adelberger's group (this experimental group will play a major role in the subsequent history) at Washington had some references on the modeling of convective effects on torsion balances and that he would send them along, which he did on 27 March. He added that the only effect contained in the references was that thermal gradients might cause a systematic drift in the equilibrium position of the balance, a point confirmed by his own work. Talmadge (14 May) later reported that there was some empirical data on the possible size of thermal effects. He noted that the $RaBr_2$ sample was known to cause thermal gradients because of the heat generated by the radioactive radium and yet had a smaller deflection than the water sample. This seemed to cast doubt on the Dicke's thermal gradient explanation of the effects.

On 18 June Talmadge received a letter from Dicke outlining his model in more detail using horizontal thermal gradients. Talmadge estimated that it would have taken Eötvös at least 150 days, and probably more, to complete his measurements. He found it implausible that such thermal gradients would have remained constant over so long a period.

These objections to Dicke's model subsequently appeared in both the Lake Louise paper and in the explicit answer to the Chu–Dicke paper (Fischbach et al. 1986e,f).

The collaborators were also in contact with a group at Fermilab that was analyzing the results of the E621 experiment on the K_S^0 lifetime in the momentum range 100–300 GeV/c. Recall that the energy dependence of the K^0–$\overline{K}^0$ parameters was one of the important pieces of evidence that had both suggested and supported the idea of a Fifth Force, so there was considerable interest among Talmadge, Fischbach, and Aronson in the results of this experiment. There was both cooperation and communication between the two groups. Talmadge and the others received data and other information from the E621 group and attempted to analyze it on their own. There were difficulties in the analysis involving the acceptances of the apparatus and Talmadge and Aronson reached no conclusion. The results of the experiment were published in 1987 (Grossman et al. 1987) and (p. 18): "The results were completely consistent with Lorentz invariance. No evidence was found for the momentum dependence suggested by the intermediate range 'fifth-force' hypothesis."

The electronic mail also documents the large amount of interest generated within the physics community concerning the Fifth Force. During the period between 6 January 1986, the publication date of the first Fifth Force paper and 22 July 1986, the end of the email record, Fischbach gave sixteen talks on the subject and Aronson fourteen.[33] There was also concern within the collaboration about how their work was being received and the electronic mail includes several comments on the fact that the talks were well received. For example, on 19 May Tobias Haas at Stony Brook wrote to Aronson: "On Wednesday we had a talk here by Dr. Fischbach and he talked about your common work. It was a very impressive lecture—I think physics at its best (even if it should turn out there is no fifth force)." There was also considerable interest in the popular press. The collaborators mentioned articles in *Time*, *Newsweek*, the *New York Times*, and *Scientific American* as well as *Science et la Vie* (France), *Corriera della Sera* (Italy), and the *Jerusalem Post* (Israel). Fischbach even appeared on CNN. It is fair to say that the hypothesis of a Fifth Force aroused substantial interest within the physics community as shown by both the publications in professional journals discussed earlier and by the very large number of talks given.[34] It also attracted significant popular interest.

Aronson's efforts included more than his work on the Fifth Force and the email record reflects this. At the time he was also working on the DO spectrometer, still under construction at Fermilab. The spectrometer includes a liquid argon calorimeter with depleted uranium plates. One question was whether or not such plates could be fabricated to within the tolerances required. Aronson made several trips to CERCA, a French firm, to examine their test plates. Although there were some problems, things seemed to be going well when a design change in the spectrometer was proposed. The original design ($N = 2$) included modules of a certain width. It was later proposed that the calorimeter consist of fewer modules, each twice as wide as originally proposed ($N = 4$). Wlodek Guryn, a physicist at

---

[33]Fischbach's talks were at TRIUMF, Stanford, Washington (Physics), Michigan, Michigan State, the National Science Foundation, Maryland, California (Berkeley), Washington (Geophysics), Cornell, Stony Brook, New York Academy of Sciences, the Lake Louise Conference, the Eleventh International Conference on Gravitation and Relativity, the Niels Bohr Institute, and the XXIII International Conference on High-Energy Physics. Aronson's talks included three at CERN, Zurich, Oxford, Louvain, Rutherford Laboratory, Heidelberg, Berlin, Paris, Annecy, Padova, Brussels, and DESY.

I suspect that at least one of the reasons for giving so many talks was the desire of the collaborators to persuade others to work on the Fifth Force. As discussed later, they had some success.

[34]Fischbach has kept an accurate record of the talks he has given on the Fifth Force. By September 1990 the number had reached 62.

Brookhaven National Laboratory, who was also working on the project broke the
news to Aronson (17 Feb):

> Yesterday we decided to go to $N = 4$ geometry. Major reason is that this matches better jet
> size and hence increases our chances for having clean events where all jets are contained
> within crackless volumes [...]. There are obvious mechanical and engineering drawbacks
> to our decision. But maybe the physics benefits are worth all the trouble. So could you let
> us know what considerations led us to $N = 2$ choice for hadronic.

Aronson did not agree with the move (18 Feb):

> Hi Wlodek; tell me this is a joke! I find it (almost) inconceivable that such drastic design
> changes are being considered at this stage [...]. You mentioned you wanted input from me
> by Monday, but I have only received this Tuesday morning. In any case I don't know what
> to say in a few words to deflect such a move; that is certainly what I would do if I could.
> It seems that all practical aspects of module construction would be more difficult. (Plate
> thickness and flatness tolerances, gap tolerances, signal board routing and connections,
> intermodular cracks get bigger, modules get heavier, etc., etc., etc., etc.)

Ultimately Aronson did not prevail and the $N = 4$ geometry was adopted in
the very nearly completed spectrometer (as of spring 1991). This is the way things
sometimes happen in large collaborations and the correspondence indicates some of
the difficulties in turning an experimental design into a working apparatus.[35]

This examination of the email has allowed us to compare the public, published
reactions to the comments and criticism, discussed in the previous section, with the
private reaction contained in the email. Allowing for some more colorful language
in the email, most of which I have not included, we see no difference between the
two records. What we have seen is the care and effort devoted to trying to do good
science. I would not wish to generalize too much from this one example, but it does
show that as far as evidential questions are concerned the public record is reliable.[36]

2.3 Is It Rising or Falling?

Before discussing the history of the experiments it is worth describing briefly the
kinds of experiments that will appear in the story. Four of these are illustrated
in Fig. 2.3. The top row shows experiments designed to look for a composition-
dependent force, one that depended on the nature of the materials used. These

---

[35]The design change resulted in CERCA being eliminated as a provider of plates because their
rolling mill was not wide enough for the wider plates required by the new design.

The email also shows other aspects of a physicist's activities. As a senior scientist in the DO
group, Aronson was also involved in the hiring of a post-doctoral research associate to work with
the group. He was asked to look around at CERN for likely candidates and was also asked for his
opinion of candidates.

[36]This is not to say that the process is not sometimes made to appear more rational than it actually
was, that the story might seem more logical and inevitable in the public record, but that the
evidential relations between experimental results and theory remain essentially the same.

Fig. 2.3 Types of experiments to measure the Fifth Force. The *upper row* shows composition-dependence experiments. The *bottom row* shows distance-dependence experiments, or tests of the inverse-square law. The *left column* shows terrestrial sources, the *right column* shows laboratory/controlled sources (From Stubbs 1990a)

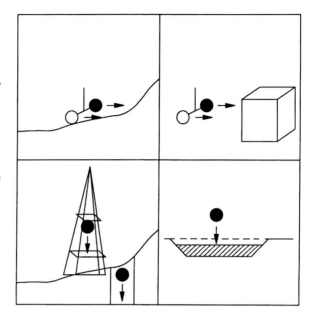

are similar to the original Eötvös experiment. As discussed earlier, one needs a local mass asymmetry to see an effect of such a short range force. These were provided by either a terrestrial source, a hillside or a cliff, or by a large, local, laboratory mass. A variant of this experiment was the float experiment, in which an object floated in a fluid and in which the difference in gravitational force on the float and the fluid would be detected by the motion of the float. These were done with terrestrial sources. An important experiment, not shown, was the repetition of Galileo's experiment of dropping two different masses and observing the difference in acceleration.

A second type of experiment looked at the distance dependence of the gravitational force to see if there was a deviation from Newton's inverse-square law. These are shown in the second row. One kind of experiment measured the variation of gravity with position, usually on a tower or in a mineshaft or borehole. The measured values of gravity were then compared with those calculated using a model of the Earth, surface gravity measurements, and Newton's law. This made accurate knowledge of the local terrain and mass distribution extremely important. A second kind of distance-dependence experiment looked at the difference in gravitational force created by a variable source, such as a lake whose water level changed.

The paleophysics era of the Fifth Force, as both Fischbach and Aronson called it, ended in early 1987 when two new experimental results were reported at the

Moriond Workshop,[37] 24–31 January, 1987 and published.[38] There was only one problem. The results disagreed.[39] The experiment performed by the Washington group, the so-called Eöt-Wash experiment, (Raab 1987; Stubbs et al. 1987) showed no evidence for a Fifth Force, while that done by Peter Thieberger (1987a,b), a physicist at Brookhaven National Laboratory, was (1987a, p. 1066) "consistent with a substance dependent, medium range force postulated in a recent analysis of the Eötvös experiment."

I will discuss these two experiments in some detail not only because they were the first new results, but more importantly because they deal with questions of possible systematic effects that will be important in later discussions. One of the most important questions in determining the validity of an experimental result is the ability to exclude background effects that might mimic or suppress a real signal.

Thieberger's experiment looked for a composition dependent force by measuring the differential acceleration between copper and water, which have different values of B/μ. The experiment was conducted near the edge of the Palisades cliff in New Jersey to enhance the effect of an intermediate range force. The experimental apparatus is shown in Fig. 2.4.

The horizontal acceleration of the copper sphere relative to the water can be detected by measuring the steady-state velocity and applying Stoke's law for motion in a resistive medium. The evacuated copper sphere was balanced with six internal counterweights to ensure that the center of mass of the sphere coincided with the center of mass of the displaced fluid. This made the apparatus insensitive to gravitational field gradients. The water temperature was kept constant at $(4.0 \pm 0.2)\,°C$, the temperature at which the density of water is a maximum, to minimize the effect of convection currents. The sensitivity of the apparatus to magnetic effects was calibrated with magnetic positioning coils. These coils produced a known nonuniform magnetic field which acted on the differing magnetic properties of water and copper producing a velocity change of $(1.4 \pm 0.1)\,$cm/h, in good agreement with the theoretically calculated value of $(1.5 \pm 0.2)\,$cm/h. This demonstrated both

---

[37]The Moriond Workshops were extremely important in the history of the Fifth Force. They were attended by many of those working on the force and provided both a formal exchange of information through the talks given, and an opportunity for informal discussion. For example, it was in these discussions that some of the experimental results were subjected to severe scrutiny and criticism, even by those whose results agreed. I attended the 1989 and 1990 workshops and heard this sort of discussion.

[38]The two papers were presented at the Moriond Workshop, 24–31 January, 1987 and were published in *Physical Review Letters* on 16 March 1987. Because the published version of the Moriond papers appeared later and was subject to later emendation I will use the *PRL* papers as the earliest results. In addition, the papers were submitted to *PRL* before the Moriond Conference, on 5 December 1986 (Thieberger) and 30 December 1986 (Stubbs). There are no major differences, although the Moriond papers contain greater detail on some points.

[39]One may speculate that the disagreement between the two results led to more work on the Fifth Force. Had the results agreed that there was no Fifth Force it might very well have settled the issue. Of course, had both experiments shown the presence of the force then further work would, no doubt, have followed.

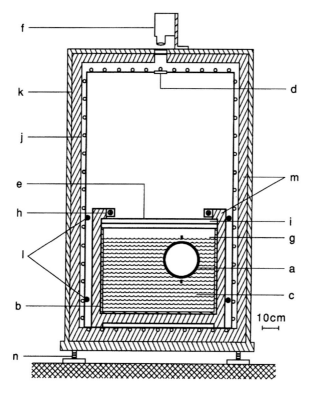

Fig. 2.4 Schematic diagram of the differential accelerometer used in Thieberger's experiment. A precisely balanced hollow copper sphere (*a*) floats in a copper-lined tank (*b*) filled with distilled water (*c*). The sphere can be viewed through windows (*d*) and (*e*) by means of a television camera (*f*). The multiple-pane window (*e*) is provided with a transparent $x - y$ coordinate grid for position determination on top with a fine copper mesh (*g*) on the bottom. The sphere is illuminated for 1 s per hour by four lamps (*h*) provided with infrared filters (*i*). Constant temperature is maintained by means of a thermostatically controlled copper shield (*j*) surrounded by a wooden box lined with Styrofoam insulation (*m*). The Mumetal shield (*k*) reduces possible effects due to magnetic field gradients and four circular coils (*l*) are used for positioning the sphere through forces due to ac-produced eddy currents, and for dc tests (From Thieberger 1987a)

the sensitivity of the apparatus to magnetic effects and showed the absence of any magnetic contaminants. The magnetic field gradient needed to produce these velocities was 10 G/m. A mu metal shield was then placed around the apparatus resulting in a residual magnetic field <0.1 G, making any magnetic effects in the actual experiment negligible.

Several checks were performed to see if there were other possible causes for the observed motions. The first test was for effects of thermal gradients. For 14 h, between points C and D in Fig. 2.5, the temperature of the west wall of the box was elevated by an average of 6 °C over the east wall of the apparatus. This was twice as large as the maximum temperature difference observed during the data

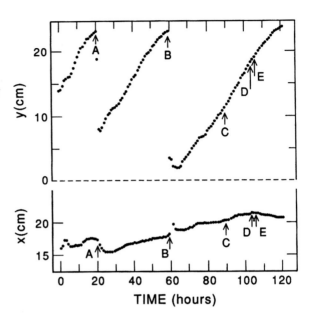

Fig. 2.5 Position of the center of the sphere as a function of time. The y axis points away from the cliff. The position of the sphere was reset at points A and B by engaging the coils shown in Fig. 2.4 (From Thieberger 1987a)

taking, and more than ten times as large as the average temperature difference. As one can see from the figure: "No appreciable effects were observed." The effect of possible leveling errors was investigated by lowering the east side of the instrument by 4.6 mm (ten times larger than the maximum leveling error) at point E in the figure. As one can see, there was no effect on the y motion, although a small effect was seen in the x motion. In order to check for possible instrumental asymmetries, Thieberger rotated the entire apparatus through 90° and found a consistent value for the velocity, (4.5 ± 0.5) mm/h normal to the cliff. The illumination frequency was also varied by a factor of 4 to look for possible heating effects. None were found. A similar measurement was made with the apparatus in another location, with no cliff, so that only small effects would be expected. The observed velocities in the x and y directions were (-0.9 ± 0.2) and (-1.2 ± 0.2) mm/h, respectively. Although these observations differ from zero, they are a factor of 4 smaller than the observed effect at the Palisades.[40] Other possible causes for the observations such as residual dipole moment and higher multipole moments, electrostatic and magnetic forces, surface tension and its temperature dependence, convection currents, vibrations, temperature gradients, and Brownian motion were also ruled out.

Thieberger's results are shown in Fig. 2.5. These measurements were taken over a five day period. There was a (4.7 ± 0.2) mm/h velocity in the y direction and (0.6 ± 0.2) mm/h in the x direction. The direction of the velocity was consistent

---

[40] Nevertheless, such a positive result might lead one to question the validity of Thieberger's result for the Palisades. With no cliff present, one expects zero velocity and a positive result might indicate the presence of systematic effects that were unaccounted for.

with the normal to the cliff, as expected.[41] The measured velocity corresponded to a difference in acceleration of $(8.5 \pm 1.3) \times 10^{-8}$ cm/s^2. Using values for α and λ, the strength and range of the presumed Fifth Force, of 0.008 and 100 m, respectively, and a physical model of the cliff,[42] Thieberger found his results consistent with the geophysical measurements. He concluded that (1987a, p. 1068): "The present results are compatible with the existence of a medium-range, substance-dependent force which is more repulsive (or less attractive) for Cu than for H_2O [...]. Much work remains before the existence of a new, substance-dependent force is conclusively demonstrated and its properties fully characterized."

The experimental apparatus for the Eöt-Wash experiment is shown in Figs. 2.6 and 2.7. It, too, was designed to search for a substance-dependent, intermediate range force, and was located on a hillside on the campus of the University of Washington in Seattle. If the hill attracted the copper and beryllium test bodies differently the torsion pendulum would experience a net torque. The pendulum, or baryon dipole, could be rotated with respect to the outside can in multiples of 90° and the entire system rotated slowly ($T_{can} \approx 6 \times 10^3$ s). If there were a differential force on the copper and beryllium one would expect to find a torque that varied with θ, the angle of the can with respect to some fixed geographical point. They detected torques by measuring shifts in the equilibrium angle of the torsion pendulum.

Here, too, great care was paid to systematic effects which might either produce a spurious signal or cancel a real signal. To minimize asymmetries the test bodies were machined to be identical within very small tolerances. Electrostatic forces were minimized by coating both the test bodies and the frame with gold, and by surrounding the torsion pendulum with a grounded copper shield. Magnetic shielding was also provided and Helmholtz coils reduced the ambient magnetic field to 10 mG. Reversing the current in the Helmholtz coils caused a_1, the signature of the interaction, to change by (3.8 ± 2.3) μrad. Scaling that result to their normal operating conditions implied a systematic error at the level of 0.1 μrad. Gravity gradients, which might result in a spurious signal if all the test bodies were not in a plane, were reduced by placing an 80 kg lead mass near the apparatus. They set an upper limit of 0.19 μrad on any possible spurious signal due to such gradients.

The most serious source of possible error was due to the "tilt" of the apparatus, which was very sensitive to such tilts. A deliberately induced tilt of 250 μrad produced a spurious a_1 signal of 20 μrad. They measured the tilt sensitivity of their apparatus carefully and corrected their data for any residual tilt. In addition, they included in their final results only those data for which the tilt was less than 25 μrad

---

[41] This was assuming the effective cliff orientation obtained by averaging over ±150 m. In a note added Thieberger noted that M.J. Good had pointed out that the Coriolis force on the sphere was not totally negligible. When this was included using the 150 m average the agreement was slightly less satisfactory. Using a ±50 m average and including the Coriolis effect actually improved the agreement. Some wag later remarked that all this experiment showed was that any sensible float wanted to leave New Jersey.

[42] Details of this model were given in Thieberger (1987b).

Fig. 2.6 Schematic view of
the University of Washington
torsion pendulum experiment.
The Helmholtz coils are not
shown (From Stubbs et al.
1987)

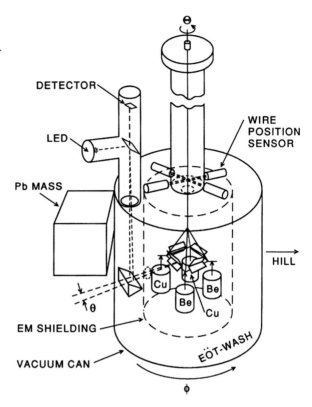

and for which the tilt correction to a_1 was less than 0.71 μrad. This was the only cut made on their data, although they noted that including all of the data gave results in good agreement with their selected sample. They also determined an upper limit of $0.11 \pm 0.0.19$ μrad due to thermal effects.

The Washington results are presented in Fig. 2.8. There is no apparent signal, although there is an offset of 4 μrad. The theoretical curves were calculated using values of α and λ of 0.001 and 100 m, respectively. The published version of the Moriond paper (Raab 1987) contains theoretical curves calculated with $\alpha = 0.01$, and show far more disagreement with the data (see Fig. 2.9). Recall that the value used by Thieberger was 0.008 and that used by Fischbach was 0.007. The *PRL* paper actually tended to understate the extent of the disagreement between the Fifth Force theory and the Washington results, although the best value of α was quite uncertain.[43]

---

[43]E. Adelberger, one of the senior members of the Washington group, remarked that the point of the *PRL* graph was to show the absence of any Fifth Force effect, even for a value of α much smaller than that needed for the geophysical or Eötvös data. He noted that, in retrospect, the *PRL* graph did not show this as well as they would have liked, so that in the Moriond paper a more realistic

Fig. 2.7 A close-up view of
the Eöt-Wash torsion
pendulum (Courtesy of Eric
Adelberger)

The Washington group concluded (Stubbs et al. 1987, p. 1072): "Our results
rule out a *unified* [emphasis added] explanation of the apparent geophysical and
Eötvös anomalies in terms of a new baryonic interaction with $10 < \lambda < 1400$ m
and make it highly improbable that the systematic effects in the Eötvös data are due
to a new fundamental interaction coupling to B [the baryon number]." They also
presented a plot (Raab 1987, p. 575) showing the limits placed on α and λ from
their experiment, the geophysical measurements of Stacey et al., and Thieberger's
experiment (see Fig. 2.10). The inconsistency is apparent.

These results were problematical for the physics community. Both experiments
appeared to be carefully done, with all the plausible and significant sources
of possible error and background adequately accounted for, and yet the two
experiments disagreed. One cannot always distinguish between a correct and an
incorrect experimental result on the basis of methodology. Neither of these two
experiments contained an obvious error.[44] The Washington experiment argued
against the presence of a Fifth Force, while Thieberger's result was consistent with
the presence of such a force.

There was very little published criticism of the two experiments, although there
was extensive private discussion. Thieberger's result, which was in disagreement
with currently accepted physics, was subjected to more scrutiny. This was not
surprising in view of the considerable evidence that already existed supporting
existing gravitational theory. Kim (1987) suggested that Thieberger's result might

---

value of α was used (private communication). At Moriond, Stacey et al. (1987a) suggested values
of α between 0.007 and 0.013, depending on the range of the force chosen.

[44]I am grateful to Jim Woodward for helpful discussions of this point.

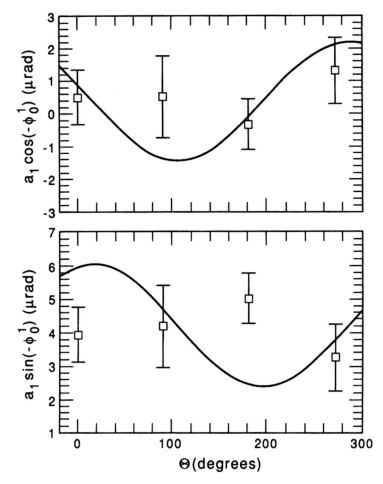

Fig. 2.8 Deflection signal as a function of θ, the variable angle of the stage. The theoretical curves correspond to the signal expected for $\alpha = 0.001$ and $\lambda = 100$ m (From Stubbs et al. 1987)

be explained by thermal convection. His model required a temperature difference of only 0.0037 °C, far smaller than the ±0.2 °C uncertainty that Thieberger claimed. Thieberger actually controlled the temperature to approximately one thousandth of a degree, a range smaller than that needed by Kim's model. The 0.2° uncertainty gave the range of temperatures (Fischbach, private communication). Kim does not seem to have known this. Kim also noted that Thieberger's measurement of a drift velocity of approximately 1 mm/h in the absence of a cliff was not inconsistent with such a model. He did not, however, regard Thieberger's attempt to magnify such

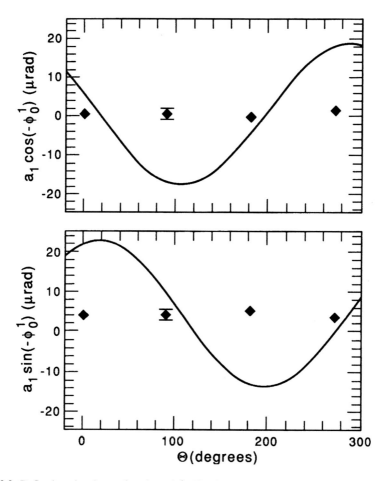

Fig. 2.9 Deflection signal as a function of θ. The theoretical curves correspond to the signal expected for $\alpha = 0.01$ and $\lambda = 100\,\mathrm{m}$ (From Raab 1987)

convection effects by heating the west wall of his apparatus as conclusive. I have found no discussion of Kim's criticism by Thieberger or by anyone else.[45]

Although the most obvious conclusion was that one of the two experiments was wrong, theorists speculated whether or not one might reconcile the two results. The starting point was a multicomponent model of the Fifth Force of the form

$$V = -\frac{Gm_1 m_2}{r}\left(1 + \sum_{k=1}^{N}\alpha_k e^{-r/\lambda_k}\right) \, .$$

---

[45] Keyser (1989) also considered a thermal convection explanation of Thieberger's results. This will be discussed below.

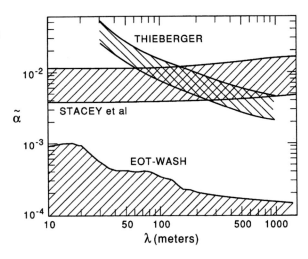

Fig. 2.10 Experimental limits on α and λ (From Raab 1987)

Recall that the original Fifth Force proposal was

$$V = -\frac{Gm_1m_2}{r}\left(1 + \alpha e^{-r/\lambda}\right) ,$$

a single component force. Such a multicomponent model had been discussed in the Gibbons and Whiting (1981) paper that had summarized the measurements of the gravitational force, and the possible anomalies. This kind of potential could arise in quantum-gravity theories that incorporated both a scalar (spin 0) and vector (spin 1) exchange particle in addition to the usual spin 2 graviton (Goldman et al. 1986). Such a potential could also result in both attractive and repulsive forces that could, under the appropriate circumstances, cancel each other out. At Moriond, Fischbach (1987) suggested that a two-component force, dependent on both $\Delta(B/\mu)$, the original Fifth Force, and $\Delta(N/\mu)$, where N was the neutron number, would have opposite signs for the Cu–Be pair used by the Washington group, and thus might account for their null result. This could remove the apparent inconsistency between the two experimental results. Stacey et al. (1987a,b) also showed that such a two-component force was consistent with their measured variations of gravity in mineshafts.

Another possible explanation was to use a single coupling to a linear combination of baryon number and lepton number. The general framework for this had been suggested by De Rujula (1986b), by Fujii (1986), and by Fischbach et al. (1986g). Hayashi and Shirafuji (1987a) had applied the formalism to the particular experiments of Thieberger and Eöt-Wash and found that the results could be made consistent using such a coupling.[46]

---

[46]A similar suggestion had also been made by Vecsernyes (1987).

This possibility was tested in another experiment by the Washington group (Adelberger et al. 1987). They considered an interaction coupling $q_5 = B \cos \theta_5 + L \sin \theta_5$, where θ_5 had a value that could explain both their previous null result for Cu–Be and also Thieberger's positive result for Cu–H$_2$O, where B and L were baryon and lepton number, respectively. They replaced the copper weights in their torsion pendulum with aluminum weights. This should have given a nonzero result. Their results, along with one standard deviation results from Thieberger are shown in Fig. 2.11. They conclude (Adelberger et al. 1987, p. 851): "There is no region where all the data are consistent at the 1σ [S.D.] level. In particular, our new results appear to rule out the possibility that our recent work and Thieberger's are both consistent with a *single* [emphasis added] Yukawa interaction coupling to any linear

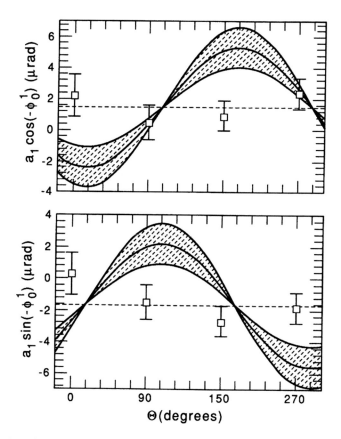

Fig. 2.11 The deflection amplitude as a function of θ. The *shaded area* shows Thieberger's $\pm 1\sigma$ error band (From Adelberger et al. 1987)

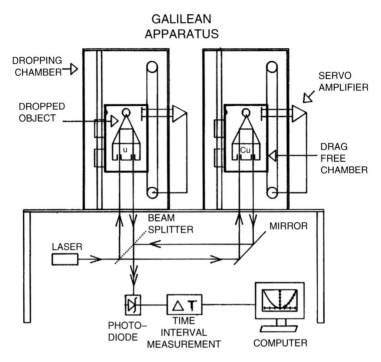

Fig. 2.12 The experimental apparatus of Niebauer et al. (1987). A repetition of the Galileo experiment

combination of B and L for $10\,\text{m} < \lambda < 1000\,\text{m}$ except for a small region around $\theta_5 = -63°$, where $q_5 \approx I_3$, the third component of isospin."[47]

The experimental situation quickly became even more complex. Niebauer et al. (1987) looked for a composition dependent Fifth Force by performing a modern day Galilean experiment. The apparatus is shown in Figs. 2.12 and 2.13. They dropped two objects of different composition, copper and depleted uranium, and found that the two accelerations were equal to within 5 parts in 10^{10} of the normal gravitational acceleration, or no composition dependence. They found a value for $\alpha\lambda = (1.6 \pm 6.0)\,\text{m}$, "which is clearly inconsistent with the value obtained in the reanalysis of the Eötvös experiment ($\alpha\lambda = 24 \pm 3$) m done by Fischbach et al." I note, however, that because of the greater uncertainty in their experiment their value for $\alpha\lambda$ was consistent with *both* the corresponding values reported by Thieberger and Eöt-Wash of (1.2 ± 0.4) m and <0.1 m, respectively.

This was not the case for the result reported by Boynton et al. (1987). Boynton, using a torsion pendulum made of aluminum and beryllium, the same materials

---

[47] The isospin of a nucleus depends on the difference between the number of protons Z and the number of neutrons N in the nucleus. Thus I_3, the third component of isospin, is $(Z - N)/2$.

Fig. 2.13 Tim Niebauer and the apparatus for the Galileo experiment (Courtesy of Jim Faller)

used in the second Washington experiment, found a positive result for the presence
of the Fifth Force. His results were, however, in disagreement with Thieberger's
result (Boynton et al. 1987, p. 1388): "Except for the intersection near $\beta = 0$,
[assuming a coupling to isospin] our result is in disagreement with Thieberger's,
although consistent with the other experiments [the two Washington experiments
and Niebauer's experiment on falling bodies]."[48]

Boynton's experiment used a torsion pendulum consisting of a ring, half of which
was aluminum and half beryllium (Fig. 2.14). The experiment was conducted in a
tunnel at the base of a 130 m high near-vertical granite wall to maximize the effect
of a composition dependent force. The site of the experiment along with various
aspects of the experimental apparatus are shown in Figs. 2.15, 2.16, 2.17, and 2.18.
If such a force existed then there would be a fractional change in the period of the
torsion pendulum $T(\theta) - T(\theta + \pi)/T$, when the pendulum was rotated by 180°.
They checked that the effects that might either mask or mimic the signature of a
composition-dependent force were negligible. These included magnetic effects on
the pendulum, thermal gradient effects, and departures of the pendant from level.
Taken together in quadrature the measured upper limit due to these effects was
5×10^{-7} in $\Delta T/T$, which was small, approximately 10 %, when compared to the
observed signal.

---

[48] β is a parameter used by Boynton and does not refer to v/c. In Boynton's terminology $\beta = 0$
corresponds to a coupling to isospin.

Fig. 2.14 Equilibrium
position of Boynton's torsion
pendulum (for $\theta = 45°$). The
dipole axis is labeled D
(From Boynton et al. 1987)

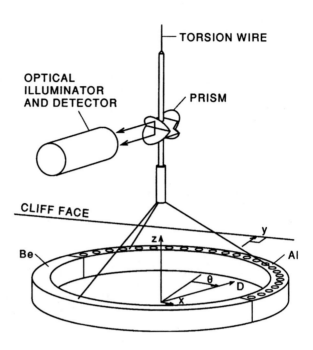

One effect that was not negligible was the well-understood coupling between gravity gradients and the tilt of the pendulum mass distribution out of the horizontal plane. They reduced this effect by tight fabrication tolerances and by placing two lead masses on opposite sides of the ring in the y-z plane. The experimenters then made use of this effect in taking their data. They first measured the ambient gravity gradient by using a solid aluminum pendulum deliberately tilted by $2°$. They then interposed the lead and again measured $\Delta T/T$. They also measured $\Delta T/T$ for the aluminum–beryllium pendulum, with and without the lead. These four measurements provided a unique decomposition of the Al–Be data into two components; one proportional to the gravity gradient and the other representing any additional interaction. The second "signal" component was

$$[\Delta T(\theta)/T]_{\text{signal}} = (-4.6 \pm 1.1) \times 10^{-6} \cos\theta + (+0.1 \pm 1.2) \times 10^{-6} \sin\theta \ .$$

The uncertainty cited was purely statistical. Combining it with estimates of systematic uncertainty raised the total uncertainty to 1.3×10^{-6}. The "signal" component, the coefficient of the cosine term was "significantly nonzero" (see Fig. 2.19). This was approximately a 3.5 S.D. effect, which is statistically significant. They observed (Boynton et al. 1987, p. 1387): "(i) It is unlikely that this signal is only a statistical fluctuation (formal probability $<10^{-3}$).[49] (ii) The signal is large compared to the

---

[49] As discussed below, Boynton later regarded this result as "marginally observed".

Fig. 2.15 The entrance to the tunnel in the cliff near Index, Washington, the site of Boynton's experiment (Courtesy of Paul Boynton)

largest identified systematic effect, the residual gravity-gradient effect shown in Fig. 2.19b, and is almost as large as the uncompensated effect. (iii) The phase of the signal ($181° \pm 17°$) is appropriate (modulo π) to a static interaction of the cliff mass with some kind of asymmetry between the Be and Al halves of the pendulum."

As an additional check, the experiment using the Al–Be pendulum was done in the sub-basement of the Physics Building at the University of Washington. Without compensating lead masses they found

$$[\Delta T/T]_{\text{obs}} = (0.8 \pm 1.5) \times 10^{-6} \cos \theta - (2.5 \pm 1.8) \times 10^{-6} \sin \theta \ .$$

This was consistent with the result expected only from gravity

$$[\Delta T/T]_{\text{grav}} = (0.9 \pm 0.5) \times 10^{-6} \cos \theta - (5.5 \pm 0.5) \times 10^{-6} \sin \theta \ .$$

Fig. 2.16 Paul Boynton
standing next to his thermally
insulated experimental
apparatus. Notice that
Boynton has also insulated
himself against the low
temperature in the tunnel
(Courtesy of Paul Boynton)

This implied that the effect observed in the cliff tunnel was valid.[50]

Geophysics also provided some evidence supporting the hypothesis of a Fifth Force. We discussed earlier some of the measurements of G, the gravitational constant, done by Stacey and his collaborators. In early 1987 Stacey and others published a review paper entitled *Geophysics and the Law of Gravity* (1987b). They noted that underground measurements of gravity all seemed to favor a value

---

[50]Further checks on the apparatus were also performed. The composition dipole axis was occasionally rotated $180°$ relative to the housing and the optics and half of each major data set was acquired in this "reversed" mode. This guarded against any effects dependent on instrument-pendulum orientation. No significant effect was seen. They also looked for correlations between the data and other observables: time of day, housing temperature, change in housing temperature, rank order of a measurement in a given day, etc. "No significant correlations are present in the data."

Fig. 2.17 Details of the
copper–polyethylene torsion
pendulum used by Boynton in
the Index III experiment
(Courtesy of Paul Boynton)

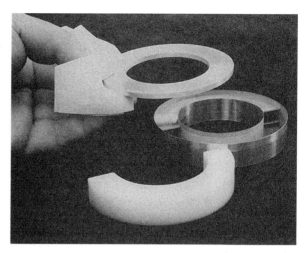

Fig. 2.18 The assembled
Index III torsion pendulum
(Courtesy of Paul Boynton)

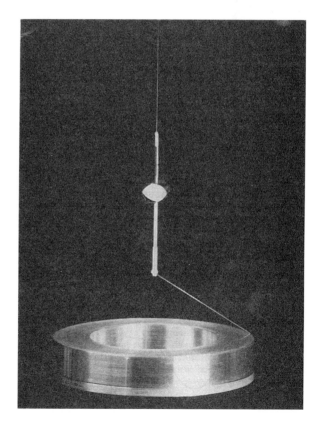

Fig. 2.19 (a) Observations
of Al–Be pendulum at 45°
increments in θ. (b)
Decomposition into
gravity-gradient and "signal"
components (From Boynton
et al. 1987)

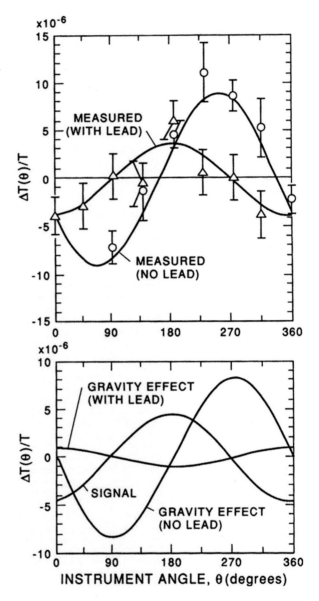

of G higher than that obtained in a laboratory (see Table 2.2). Figure 2.20 shows the difference between the measured and calculated values of gravity for their Hilton mine data, along with curves calculated for different values of the Fifth Force. They stated (Stacey et al. 1987b, p. 157): "The evidence is still less than completely conclusive but it has now become difficult to find explanations of the geophysical observations other than non-Newtonian gravity." They also felt that the care documented in their previous work (Holding and Tuck 1984; Holding et al.

Table 2.2 Values of G from underground measurements. The quoted uncertainties are statistical. The best laboratory value for G was then $6.6726(5) \times 10^{-11} \, \mathrm{m^3 \, kg^{-1} \, s^{-2}}$ (From Stacey et al. 1987b)

Reference	$G[10^{-11} \, \mathrm{m^3 \, kg^{-1} \, s^{-2}}]$
Whetton et al. (1957) (mine)	6.795 ± 0.021
McCulloh (1965) (mine)	6.733 ± 0.004
Hinze et al. (1978) (borehole)	6.81 ± 0.07
Hussain et al. (1981) (mine)	6.705 ± 0.016
Stacey et al. (1987b)	6.720 ± 0.024
(Hilton mine)	
(Mount Isa mine)	$6.704^{+0.089}_{-0.025}$

Fig. 2.20 Plot of the differences between measured gravity and calculated values, assuming Newton's law and the laboratory value of G. The two curves are calculated for different values of α and λ. The *solid curve* is for $\alpha = -0.0077$ and $\lambda = 200 \, \mathrm{m}$ (From Stacey et al. 1987b)

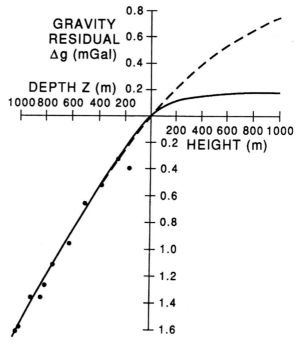

1986), particularly on local density measurements, "makes the high values of G look robust." They also remarked positively on the reanalysis of the Eötvös data by Fischbach et al. (1986a). After noting that this original work had aroused considerable comment they remarked (1987b, p. 171): "Much of the comment is strongly critical, but it is clear that the essential point made by Fischbach et al. survives the criticism." As we have seen, not everyone agreed with that judgment.

Hsui (1987) also reported a result which was inconclusive, but consistent with the results of Stacey et al. (1987b) ($G = 6.69 \pm 0.07 \times 10^{-11} \, \mathrm{m^3 \, kg^{-1} s^{-2}}$).[51]

---

[51] Interestingly, *Science* magazine reported that Hsui's results confirmed the existence of the Fifth Force (Science **237**, 819). A later letter by Zumberge and Parker (1987) pointed out that because of the large uncertainty Hsui's result was indeed inconclusive.

Although at the end of 1987, the evidential situation with respect to the existence of a Fifth Force was uncertain, to say the least, there were promises of more experimental help on the way. As early as the January 1987 Moriond workshop, three new proposed experiments were discussed. Bizzeti (1987) presented preliminary data on an experiment similar to that of Thieberger. The apparatus included a sphere floating in a liquid of the same density, but with different B/μ, located on a hillside. His preliminary data on the stability of the sphere gave an average drift of ≈ 0.04 mm/h. This was both smaller than the velocity expected for the Fifth Force under the experimental conditions and also smaller than the uncertainty of 0.2 mm/h reported by Thieberger. Newman (1987) presented plans for a torsion balance experiment, similar to the Eöt-Wash experiment, that would be capable of a precision comparable to that of the Washington group, and thus provide an independent check on its results. Kuroda (1987) discussed the possibility of an experiment to detect composition dependence by measuring the differential acceleration of two falling objects of different materials. This was the famous Galileo experiment that had also been performed recently by Faller and collaborators (Niebauer et al. 1987).

Thus, three of the early experiments were to be repeated. These would not be Heraclitean repetitions, but they would be experiments using similar apparatuses based on the same physical principles.[52] With good luck they would help to resolve the uncertainty.

There were also other suggestions for experiments. Silverman (1987a,b) and Nobili et al. (1987) suggested orbiting space experiments that they believed would provide more sensitivity. Hayashi (1987) showed, however, that not only would Silverman's proposed experiments not be feasible for practical reasons, but also that, in general, space experiments would not be able to detect the Fifth Force. Pusch (1987) discussed some of the experimental problems of Eötvös-type experiments and suggested a hyperforce resonance detector, which he believed would be better. Hayashi and Shirafuji (1987b) proposed improvements in the Kreuzer experiment and also estimated the size of the possible measurable effects. Goldman et al. (1987a,b) calculated the differences in force between matter and antimatter. They found possible effects large enough for them to urge completion of the proposed experiment comparing the fall of protons and antiprotons.

Theoretical work was also continuing. This work took two forms: (1) looking at the implications of the hypothesized Fifth Force in other areas and (2) attempting to provide an explanation of the Fifth Force.

One area where scientists looked for possible observable effects of the Fifth Force was in stellar structure. Gilliland and Dappen (1987) investigated this and found that the changes to solar structure, neutrino fluxes, and oscillation frequencies were within existing observational and theoretical limits, but that the modifications to stellar lifetimes were large enough to merit investigation, should the Fifth Force

---

[52]For a discussion of the confirmation provided by the "same" and "different" experiments see Franklin and Howson (1984).

be shown to exist. Glass and Szamosi (1987) also looked at stellar structure and found that for the then currently accepted values for $\alpha\lambda$ of approximately 1, the observable effects would be negligible (approximately one part in a million). For larger values of either α or λ, which could be accommodated within existing Fifth Force phenomenology, the effects might be comparable in size to those due to general relativity. D'Olivo and Ryan (1987) found that a Fifth Force would have no observable effects on a Newtonian cosmology.[53]

The search for a deeper explanation of the Fifth Force was also proceeding. Peccei et al. (1987) looked at a scalar particle solution to the problem of the vanishing cosmological constant. They found that their solution, the cosmon, resembled the Fifth Force. It's strength was approximately 1 % of gravity and its range was $\leqslant 10^4$ m. The conflicting data from the Washington and Thieberger experiments also complicated their analysis. They eagerly awaited new experimental results. Bars and Visser (1987) continued their work on the Fifth Force as evidence for higher dimensions, but Cho (1987) showed that such an effect was possible for only a very limited class of theories. Nieto et al. (1987a,b,c) continued to look at the consequences of spin 1 and spin 0 partners of the spin 2 graviton, as a possible source of the Fifth Force. Moffat (1986) found that his nonsymmetric theory of gravity did give rise to a composition dependent, intermediate range force. Thus, there were several possible explanations of the Fifth Force, but none of them had the support of a significant segment of even those working on the problem, much less of the physics community as a whole.

Some of the new experimental results that Peccei, Sola, and Wetterich had hoped for soon arrived, and others were on the way. At the second annual meeting of the Fifth Force faithful, the Moriond Workshop, 23–30 January 1988, four new experimental results were presented. Unfortunately, rather than clarifying the situation, they added to the confusion.[54]

Eckhardt et al. (1987a) presented results from a new type of experiment.[55] They measured the acceleration due to gravity at various heights on a 600 m tower and compared them with the values calculated from an upward continuation based on ground measurements (see Fig. 2.21). They found that (p. 575): "A significant departure from the inverse-square law was detected, asymptotically approaching $-500 \pm 35\,\mu\mathrm{Gal}(1\,\mu\mathrm{Gal} = 10^{-8}\,\mathrm{ms}^{-2})$ at the top of the tower; this indicates that at the base of the tower there is a non-Newtonian attractive force that falls

---

[53]Kuhn and Kruglyak (1987) looked at modifications of Newton's law at planetary and cosmological distances and found them to be consistent with existing observational constraints. This was not the Fifth Force, the distance scale was much larger, but the authors noted that it was suggested to them by work on that force.

[54]All of these results were eventually published in journals. In the case of Eckhardt et al. (1987b), Stubbs et al. (1989b), and Thomas and Vogel (1990) there were no significant changes. For Bizzeti et al. (1989b) considerably more data was included although their general conclusion did not change.

[55]These results were first presented at the December 1987 meeting of the American Geophysical Union.

Fig. 2.21 Eckhardt's
experimental results fitted to
a scalar Yukawa model (From
Fairbank 1988)

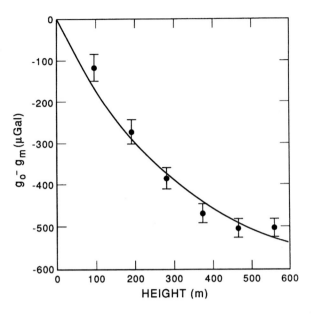

Fig. 2.21 Eckhardt's experimental results fitted to a scalar Yukawa model (From Fairbank 1988)

off rapidly with elevation. The results are marginally consistent with a one term Yukawa type attractive force, but they are fully consistent with two Yukawa type forces, attractive and repulsive [...]." This positive result was supportive of the existence of a Fifth Force, but it was also inconsistent with the repulsive force required by the geophysical measurements of Stacey et al.[56] For a single Yukawa type force they found $\alpha = 0.0204$, $\lambda = 311$ m, whereas Stacey et al. had found $\alpha = -0.0075 \pm 0.0038$, with a range of approximately 200 m. The agreement with a two-Yukawa model led some to speculate that there was not only a Fifth, but also a Sixth, Force.

The second result, reported by Thomas et al. (1988), analyzed gravity and density data from five boreholes at the Nevada Test Site. This, as did Stacey's results, involved a downward rather than an upward continuation in their calculation of gravity. They noted later (1990) that a downward continuation was more sensitive to noise in the surface gravity data. They found a 2.5 % discrepancy between the observed gravity gradient and that predicted by a standard Newtonian model of the Earth. This disagreed in magnitude with Stacey's 0.52 % discrepancy, and in both sign and magnitude with Eckhardt's 0.29 % discrepancy. They also noted, however, that the measured free air gradients disagreed with those calculated from the model, and they concluded that (p. 591) "the model does not reflect the total

---

[56]Eckhardt also quoted Airy (1856, p. 299) on the difficulties of gravity measurements: "We were raising the lower pendulum up the South Shaft for the purpose of interchanging the two pendulums, when (from causes of which we are yet ignorant) the straw in which the pendulum-box was packed took fire, lashings burnt away, and the pendulum with some other apparatus fell to the bottom. This terminated our operations of 1826."

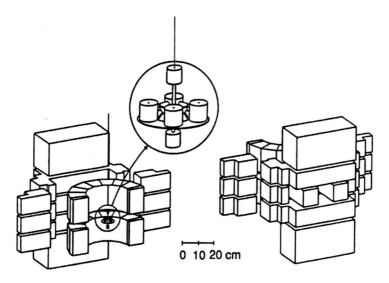

Fig. 2.22 The New Washington apparatus using a local lead source (From Adelberger et al. 1988)

mass distribution of the Earth with sufficient accuracy to make a statement about Newtonian gravity [or about the Fifth Force]." Although this result did not help to resolve the issue of the existence of the Fifth Force it did raise the important question of the adequacy of the model of the Earth used in the calculations of gravity, an important point in the subsequent history.

The Washington group (Adelberger et al. 1988) also presented new results. Recall that the earlier experiments did not rule out an interaction having $\theta_5 \approx 63°$, or a coupling to $N - Z$, where N is the number of neutrons in the nucleus, and Z the number of protons. The previous experiments had used terrestrial sources for which $N - Z$ was approximately 0. The new Washington apparatus (shown in Fig. 2.22) used a large (800 kg), local, lead mass to provide a source for the possible force. Lead has a $(N - Z)$/volume approximately 120 times that of, for example, the Index cliff used by Boynton, and so that even a small lead source could have a strength comparable to terrestrial sources of much larger mass. The Washington results, along with the Fifth Force predictions for $\alpha_5(\theta_5 = 63°) = 3.5 \times 10^{-2}$ and $\lambda > 1$ m are shown in Fig. 2.23. The value of α_5 was chosen to agree with Boynton's results. There is clear disagreement with the predictions. Their results at the 1 S.D. (standard deviation) level were inconsistent with Boynton's for $N - Z$ coupling and $\lambda \leqslant 1000$ m. For 2 S.D. the disagreement was for $\lambda \leqslant 500$ m. They concluded that there was not yet any good evidence for a composition-dependent force.

Bizzeti et al. (1988) used a floating body experiment, which was very similar to that of Thieberger. Their results, however, differed dramatically from his. These results, taken for a period of 15 days, showed that the floating body was remarkably stable (see Fig. 2.24 and Table 2.3). They found a drift in the East–West direction of approximately 3 μm/h in disagreement with the Fifth Force prediction

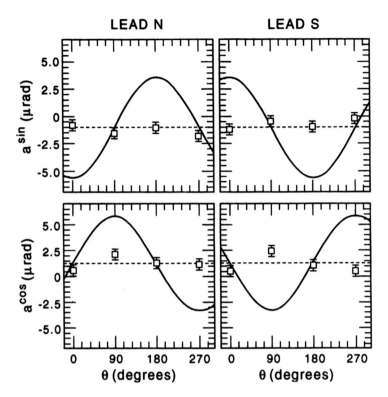

Fig. 2.23 The Washington results along with the predictions of the Fifth Force for $\alpha = 0.035$ and $\lambda > 1$ m (From Adelberger et al. 1988). The value of α was chosen to be consistent with Boynton et al. (1987) results

of 130–270 μm/h in the West–East direction. In the North–South direction where no drift due to the Fifth Force was expected the average velocity was approximately 2 μm/h. This measured the stability of the sphere. Thus, the velocities in the two perpendicular directions were approximately equal and far smaller than that predicted by the Fifth Force. Tests of possible thermal or gravity gradient background effects were being planned, although the experimenters did not regard them as plausible explanations of the stability of the float. Worries were also expressed that the density gradients used in this experiment might reduce the motion of the float.

Thus, the experimental situation seemed even more confused than it had been. Boynton, Adelberger, and Bizzeti seemed to have ruled out Thieberger, but Boynton and Adelberger themselves disagreed. There was also a disagreement in the sign of the effect between Stacey's mine measurements and Eckhardt's tower result.

It was, however, pointed out by Talmadge and Fischbach (1988) that there was a very small region where a single component Fifth Force, with almost pure baryon coupling and with $\lambda \simeq 1000$ m, that was compatible with all of the experimental results. This was precisely the region predicted by the cosmon model of Peccei,

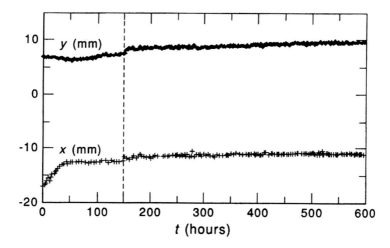

Fig. 2.24 Position of the sphere completely immersed in liquid as a function of time. The *vertical line* marks the time at which the restraining wires were removed (From Bizzeti et al. 1988)

Table 2.3 Average velocities in the E → W and in the N → S direction. Errors shown are only statistical. From a Fifth Force one expected a drift velocity in the range 130–270 μm/h, directed approximately from West to East (From Bizzeti et al. 1988)

Sphere	Time interval (h)	$v_{E \to W}$ (μm/h)	$v_{N \to S}$ (μm/h)
At the surface	144–209	-2.2 ± 1.3	3.2 ± 1.3
	209–282	0.6 ± 1.3	3.4 ± 1.3
Immersed	196–269	3.4 ± 2.6	-1.9 ± 2.9
	269–336	1.2 ± 1.6	2.8 ± 1.7
	336–407	3.3 ± 1.1	2.8 ± 1.1
	407–485	3.2 ± 1.1	2.9 ± 1.3
	485–555	2.2 ± 1.2	4.8 ± 1.2
	555–625	4.2 ± 1.1	2.9 ± 1.1

Sola, and Wetterich in a paper presented at the workshop by Wetterich (1988). Fujii (1988) presented a two component, scalar–vector model of the Fifth Force that could reconcile all of the experimental results except those of Eckhardt, which were unavailable when he did his calculations, but which could possibly even reconcile the Stacey–Eckhardt disagreement. This was also true for the scalar–vector model presented by Hughes et al. (1988).

The evidential confusion concerning the Fifth Force could be made consistent, but only at the price of either very restricted values of the parameters, for which no plausible explanation or reasons existed, or of an increasingly complex theory.

As Fischbach et al. remarked (1988, p. 72): "The multicomponent scenarios [they had three in their latest model] depend for their viability on various assumptions regarding the strengths and ranges of the different contributions. At the present stage these relations are merely assumed as needed on phenomenological grounds [i.e., to

fit the experimental results]. However, should any of these scenarios remain viable
as additional data become available, then a deeper understanding of such models
will be called for."

2.4   The Force Is Falling

The 1988 Moriond Workshop marked the high point of evidence in support of
the Fifth Force, or, perhaps more accurately, the point of maximum evidential
confusion. Although a few later experimental results would be compatible with
the existence of such a force, they would also be compatible with alternative
explanations that did not include it. Most of the further measurements would confirm
Newtonian gravity, and set more stringent limits on the presence of the Fifth Force.
In addition, doubts would be cast on some, but not all, of the earlier positive
results.[57]

The first of the new experimental results to appear was that of Fitch et al.
(1988). Their experiment was similar to that of the Washington group and used
a torsion balance located on a steep hillside. To look for a possible composition
dependent effect they used a balance of copper and polyethylene, which was similar
in composition to the copper–water pair used by Thieberger. They found, for both
a coupling to baryon number $(N + Z)$ or to isospin $(N - Z)$, values for $\alpha\lambda$ that
agreed within experimental uncertainty and which were $\alpha = -0.04 \pm 0.07$ m for
$25 < \lambda < 400$ m increasing linearly to -0.05 ± 0.09 m at $\lambda = 1.6$ km. This should
be compared to the value of 1.2 ± 0.4 m obtained by Thieberger. Their result was
consistent with Newtonian gravity and with the torsion balance results of the Eöt-
Wash group. They pointed out, however, that their measurement was not sufficiently
sensitive to either confirm or refute Boynton's result.

A further result consistent with Newtonian gravity was found by Stacey and his
collaborators (Moore et al. 1988). They measured the gravitational force exerted
by layers of lake water on steel masses suspended at different levels in the lake by
measuring the difference in force on those masses as the water level in the pumped-
storage lake changed. They found, for an average separation of 22 m, a value for
$G = (6.689 \pm 0.057) \times 10^{-11}$ m^3 kg^{-1} s^{-2} in agreement with the best laboratory
value of $(6.6726 \pm 0.0005) \times 10^{-11}$. Because of the large experimental uncertainty,
this result was also consistent with the then current estimates of the strength and
range of the Fifth Force.

At the 1988 Grossmann Meeting (Blair and Buckingham 1989), Kuroda and
Mio (1989a) presented data from another modern repetition of Galileo's falling
body experiment. They found, for mass pairs of Al–Cu and Al–C, differences in
gravitational acceleration $\Delta a = (-0.13 \pm 0.78)\,\mu$Gal and $(-0.18 \pm 1.38)\,\mu$Gal,

---

[57]Shortly after the 1988 workshop the detailed renalysis of the Eötvös experiment (Fischbach et al.
1988) appeared. This paper also discussed the existing evidential uncertainty.

respectively. By comparison, using the existing estimates from the positive Fifth Force results, for the strength and range of the Fifth Force gave predictions of 1.8 and 3.0 μGal, respectively, in disagreement with the measurements.[58]

The trend of experimental results against the existence of the Fifth Force continued, although none of these new experimental results set more stringent limits on the presence of the Fifth Force. Speake and Quinn (1988), using a beam balance with lead and carbon masses, obtained a value of $\xi = (0.8 \pm 2.0) \times 10^{-2}$. (Note that ξ differs slightly from α.[59]) This result did not supersede in precision any of the previous results arguing against the Fifth Force, but it did provide (p. 1343) "a completely independent method of verifying results from torsion balance experiments."

Such experiments had further confirmation when Cowsik et al. (1988) used a lead–copper pendulum with a laboratory lead source and found $\xi < 3 \times 10^{-3}$ for all $\lambda > 3$ m. Their actual result was $\xi = (-0.03 \pm 1.5) \times 10^{-3}$, but they cautiously set a 2 S.D. upper limit. They remarked that this was larger than the limit of $\xi < 4 \times 10^{-4}$ obtained by Adelberger, but that it helped to set limits for small values of $\lambda (< 10$ m) where hillside experiments were less sensitive.

The only new 1988 result supporting the Fifth Force was reported at the Grossmann conference. The Greenland group measured gravity in a 2 km deep borehole located in the Greenland ice cap. They found an unexplained difference between the measurements at 213 and 1673 m of 3.87 mGal (this value is taken from their later published report (Ander et al. 1989)). The anomaly was both larger and opposite in sign to that reported by Stacey et al. Once again this anomaly depended strongly on the model used to calculate the predicted Newtonian result. In fact, this result had been presented earlier in August at a press conference, and had attracted attention in the popular press. (See, for example, *Time Magazine*, 15 August 1988, p. 67.) At the Grossmann conference an informal session was held to discuss this new result. It was subjected to rather severe criticism, particularly for the paucity of good surface gravity data near the location of their measurement and for the inadequacy of their theoretical model of the Earth. It was pointed out that

---

[58] At the Grossmann conference Eckhardt et al. (1989b) presented data similar to those presented at Moriond, although the value for the discrepancy had changed slightly to (-547 ± 36) μGal. Thomas et al. (1989) presented data from both boreholes and a tower at the Nevada Test Site that showed gravitational anomalies. Once again they attributed this to defects in their theoretical model and urged great care in the use of such models, particularly in the acquisition of sufficient surface gravity measurements.

[59] Recall that α was defined by the equation

$$V = -Gm_1 m_2 / r (1 + \alpha e^{-r/\lambda}),$$

while ξ was defined by

$$V = -Gm_1 m_2 / r (1 - \xi q_1 q_2 e^{-r/\lambda}),$$

where $q = \cos \theta_5 (B/\mu) + \sin \theta_5 (2Iz/\mu)$. (B/μ) is very close to 1 for all substances, so for $\theta_5 = 0, \alpha \approx -\xi$.

there were underground features in Greenland of the type that could produce such anomalies (E. Fischbach, private communication) (Fischbach chaired this informal session).[60] This result was also presented at the fall meeting of the American Geophysical Union (Zumberge et al. 1988), where the group pointed out that it could be interpreted either as evidence for non-Newtonian gravity or explained by local density variations.[61] At this meeting, Parker (1988), a member of the Greenland group, also noted that the Stacey and Eckhardt results for non-Newtonian gravity might also be interpreted in terms of local density variations. In a post-deadline paper presented at the 1988 AGU meeting, Bartlett and Tew suggested that both the Stacey and Eckhardt positive results on the Fifth Force came largely from an inadequate modeling of the local topography.

During this period of time theoretical work continued along the same lines that it had previously—continued work on scalar–vector theories, looking at possible implications of the Fifth Force, etc.—but with no new major insights or explanations. The experimental program had become virtually independent of theory and had acquired a life of its own. In fact, except for the original suggestion of the Fifth Force, experiment had proceeded largely on its own. Rarely was any explicit theory mentioned in experimental papers. Even the suggestion of other possible parameters to explore, such as isospin or lepton number, had been either in the way of phenomenological suggestions or suggested by the experimental results.

The evidence against the Fifth Force continued to accumulate and, beginning in early 1989, at an increasing rate. To mix a metaphor, the Fifth Force was being accelerated downward. In January 1989, Bennett (1989a) reported a measurement of the difference in the force exerted on copper and lead masses by a known mass of water, located nearby. He used a Cu–Pb torsion balance located near the Little Goose Lock on the Snake River in eastern Washington, in which the water level was changed periodically to allow the passage of boats. The copper–lead comparison was chosen to maximize the sensitivity to isospin, the parameter that Boynton's work had indicated was the relevant parameter. The lock was chosen so that possible background effects could be minimized. The lock had a very large change in water level, which made the mass of water large, its structure allowed the experimental apparatus to be placed close to the water's edge, and its structure provided shielding from both the Sun and wind, which minimized thermal and mechanical effects which might mimic a possible difference in force.

The difficulties of real as opposed to ideal experiments was clearly illustrated (Bennett 1989a, p. 366): "Because the data were taken during a dry period (August 1988), separate lock fillings could not be made just for the experiment. On average there were four 'lockages' a day from barge traffic which could occur at any hour of the day or night with only a half-hour advance notice." The apparatus needed minor adjustment every 4 or 5 h and then took about 2 h to stabilize, allowing good data to be taken during the next 2 or 3 h (p. 367): "The success of a particular

---

[60]The Greenland result was not included in the published conference proceedings.

[61] This result was also published later, as discussed below.

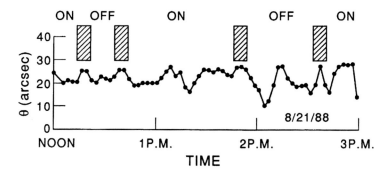

Fig. 2.25 Mean angular deviation θ of the mass dipole from parallelism to the lock. *Shaded areas* denote lock fill and drain periods. *ON* denoted lock full (From Bennett 1989a)

Table 2.4 Results for successive lock transitions (From Bennett 1989a)

Number of measurements	Deflection (arcsec)[b]	Cu–Pb acceleration toward water (10^{-8}cm/s^2)
12[a]	-0.74 ± 1.72	-0.40 ± 0.99
12	-1.60 ± 1.31	-0.86 ± 0.70
28	$+4.23 \pm 5.09$	$+2.27 \pm 2.74$
14[a]	$+3.52 \pm 6.51$	$+1.89 \pm 3.50$
12	$+10.15 \pm 5.81$	$+5.46 \pm 3.12$
12	$+4.66 \pm 5.08$	$+2.50 \pm 2.73$
26	-2.24 ± 6.65	-1.20 ± 3.57
Weighted mean	-0.47 ± 0.96	-0.25 ± 0.52

[a]No boats in lock
[b]Positive deflection means Cu rotated toward water

run depended on coincidence of this observation period with the arrival of lock traffic and, typically only one could be observed in a period of about 6 h during weekdays. Fortunately, traffic on weekends was heavier because of pleasure craft. Although consistent with individual isolated experiments, by far the best data were obtained on Sunday, 21 August 1988, when an armada of such small craft went up and down the river." A sample of the data obtained is shown in Fig. 2.25, and the experimental results given in Table 2.4. Bennett found, for a range $\lambda = 100\,\text{m}$, $\alpha = (-0.52 \pm 1.04) \times 10^{-3}$ for pure isospin coupling, in good agreement with the negative results of Adelberger and Fitch.[62]

The 1989 Moriond Workshop provided even more evidence against the Fifth Force. Not only were several very precise negative results presented, but doubts began to surface about some of the earlier positive results.

---

[62]Long (1989) raised a question concerning the possible effect of the tilt of the apparatus on Bennett's result. Bennett (1989b) argued that the effect proposed was, in fact, far smaller than his experimental uncertainty.

Newman et al. (1989; Nelson et al. 1990)[63] presented a new torsion balance result using a copper–lead balance and a lead laboratory source. As discussed earlier, such an experiment was very sensitive to an isospin coupling. The final result was $\xi = (5.7 \pm 6.3) \times 10^{-5}$ for pure isospin coupling and $\xi = (1.2 \pm 1.3) \times 10^{-3}$ for baryon coupling. This was the most stringent limit yet. The experiment was very carefully done with the experimental uncertainties introduced by magnetic coupling, tilt, Newtonian gravitational coupling, thermal effects, suspension asymmetry, and electronic asymmetry all determined. In addition, Newman et al. (1989) introduced a new check on their result. To guard against any possible experimenter bias they added an unknown (to the experimenters) quantity to their result while final data selection and analysis were being done. Only after the final result was obtained was that unknown quantity subtracted.[64]

The Washington group also presented data from both their hillside and laboratory source experiments, using an improved apparatus. Stubbs et al. (1989a,b) reported on the laboratory results. They concluded (1989b, p. 609): "Our null results rule out (at 2σ [standard deviation]) the possibility that all previous composition-dependence results could be due to a force coupling predominantly to $B - 2L$ [isospin] with a range $\lambda < 1000$ m." Their limits were, for $\lambda \geqslant 1$ m, $\alpha = (-0.14 \pm 1.24) \times 10^{-3}$ and $\alpha = (0.21 \pm 1.90) \times 10^{-3}$ for isospin and baryon coupling, respectively. They also set limits on a two-Yukawa fit to their data.

Adelberger (1989) reported on the Washington hillside experiment, which claimed a factor of 25 increase in sensitivity over their earlier work. They found, for a beryllium–aluminum pendulum, $\Delta a = (4.5 \pm 4.4) \times 10^{-11}$ cm/s^2 (p. 494): "Our upper limits lie so far below the claimed positive effects of Thieberger and Boynton et al. that we find it hard to believe that there is *any* [emphasis in original] credible evidence for a composition-dependent fifth force."[65] He also presented a graph which added the Washington results (Eöt-Wash I and III) to Fischbach's reanalysis of the Eötvös data (p. 497), "the plot which motivated so many of us to undertake a search for a 'fifth force' [see Fig. 2.26]. It is now impossible for me to believe the striking Eötvös anomaly has anything to do with fundamental physics."

Adelberger's conclusions are quite interesting. He argued quite strongly against the existence of a Fifth Force, at the level of the best experiments, and that complex model building was therefore fruitless. He nevertheless argued that the search should continue because it was a relatively cost-effective way of searching for very interesting physics.

---

[63]In most cases the papers presented at the Moriond workshop were published elsewhere later. I will give both references initially, and if there are any significant changes I will discuss them when the second paper was published, which was the time the information became generally available to the physics community.

[64]Such a bias is not unheard of and in a recent experiment the same check was made. It indicated that such a bias may very well have been present. See Franklin (1986, p. 170).

[65]Adelberger noted that the precision of their result on the equality of fall toward the Earth now matched the precision of the Roll, Krotkov, Dicke result on the equality of fall toward the Sun.

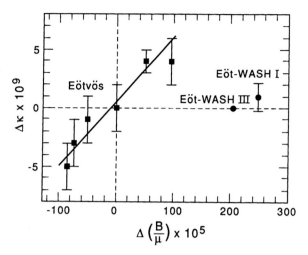

Fig. 2.26 Comparison of the Eötvös reanalysis of Fischbach et al. (1986a) with the results of the Eöt-Wash I and III experiments. The error bar on the Eöt-Wash III datum is smaller than the dot (From Adelberger 1989)

Bizzeti et al. (1989a,b) presented further results from their float experiment. They continued to find no evidence for a Fifth Force, and their limits for pure baryon coupling were compatible with those obtained by Fitch and by the Eöt-Wash group. They found a drift velocity $v < 10\,\mu$m/h, in comparison with the Fifth Force prediction of about 40–80 μm/h. Because of the substances used, they could not set any useful limits on the isospin coupling suggested by Boynton. They were, however, able to set some limits on the cosmon model of Peccei, Sola, and Wetterich.

Evidence was also presented from tower gravity experiments. The Livermore group (Kasameyer et al. 1989) presented a definite result from their gravity measurements at the 454 m high BREN tower at the Nevada Test Site. Recall that they had previously questioned the anomalies they had found because they felt that their surface gravity survey and model of the Earth was inadequate (these were the papers of Thomas et al. 1988). They now had an extensive ground survey—their own measurements at 91 stations within 2.5 km of the tower, supplemented with 60,000 surface gravity measurements within 300 km of the tower done by others. They found preliminary results in agreement with Newtonian gravity to within 93 ± 95 μGal at the top of the tower (454 m), and in disagreement with the 500 μGal discrepancy reported previously by Eckhardt at 562 m. A somewhat later paper (Thomas et al. 1989) lowered the value to $(-60 \pm 90)\,\mu$Gal. (I note that the sign of the discrepancy did not change. Kasameyer gave only the absolute value of the discrepancy.)

At the workshop, doubts were also raised concerning some of the positive Fifth Force results. Bartlett and Tew (1989a) gave more details of their earlier suggestion that the positive results found for non-Newtonian gravity by Eckhardt in 1988 might be due, in large part, to a failure to properly take into account the local terrain. They also suggested that this problem might also be present in the mineshaft data of Stacey et al. (1987a,b). They admitted that the question of whether or not the theoretical models properly accounted for local terrain was still open, and could be answered only by the experimenters themselves. They did, however, present a calculation arguing that 60–65 % of the tower residuals of Eckhardt et al. (1987b) could be explained by local terrain. A later result (1990) argued that three quarters of the anomaly could be attributed to local terrain, leaving only a small amount for terrain effects, systematic measurement error, or a Fifth Force (Fig. 2.27).

Eckhardt et al. (1989a) disagreed. They presented a preliminary reanalysis of their previous result which reduced their anomaly to $350 \pm 110 \mu$Gal. They had, since their 1988 result, been searching for possible errors and had both increased their surface gravity survey and refined their calculations (p. 526): "We also had the help of critics who found our claims outrageous."[66] These included Bartlett and Tew. Eckhardt et al. (1989a) remarked that although they disagreed with Bartlett and Tew about whether or not they had adequately accounted for the local terrain, the criticism had caused them to look more carefully at possible elevation sampling biases. They had, indeed, found one and that accounted for their revised result (p. 526): "Nevertheless the experiment and its reanalysis are still incomplete and we are not prepared to offer a final result." Their anomaly had, however, become smaller.

Boynton and Peters (1989) told the conference of a subtle problem with their apparatus. After making improvements to increase the stability and lessen

Fig. 2.27 Tower gravity residuals versus height z compared to predictions from bias in location of ground stations. *Circles* are Eckhardt's data and the *curves* use slightly different approximations to the local terrain (From Bartlett and Tew 1990)

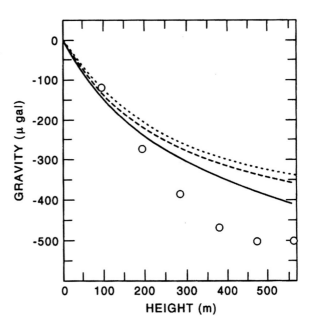

---

[66]The critics were right. In 1990, Jekeli, Eckhardt, and Romaides completely withdrew their claim of the observation of non-Newtonian gravity.

background effects they had found that their new apparatus was sensitive to a Coriolis effect that made their data unreliable. They noted, however, that this effect was due to the change in their apparatus, and was neither expected nor observed in their earlier positive result. They still had no explanation of that (p. 508) "effect that was marginally observed in that series of experiments." The apparatus had subsequently been modified to eliminate the Coriolis effect and a new series of measurements was under way. This did not cast doubt on their earlier positive result, but it did emphasize the sensitivity of the apparatus to small systematic effects. This sensitivity also applied, of course, to the experiments that argued against the Fifth Force, making the repetitions under different circumstances and with different materials more significant. The stability and consistency of the results under different conditions argued against their being an artifact. Different experiments have differing sources of error and background. (See Franklin (1986, Chaps. 4 and 6) for details of the argument.)

Not everyone was willing to take Adelberger's advice about the fruitlessness of further theoretical modeling. At Moriond, Fischbach et al. (1989) presented a new model of the Fifth Force. This was an exponential potential rather than a Yukawa type. (An exponential potential is proportional to $e^{-r/\lambda}$ whereas a Yukawa potential is proportional to $e^{-r/\lambda}/r$.) Such a model could arise if there were two interfering Yukawa potentials, such as those suggested by Eckhardt et al.'s 1988 results. Fischbach also noted that such a potential arose quite naturally in a broad class of models. One advantage of the exponential was that it offered a way of possibly reconciling the existing experimental evidence. It also lessened the importance of laboratory, as opposed to hillside, experiments. Fischbach took a much more positive view of the evidential situation than did most of those working on the subject. He argued that none of the positive results had yet been explained in terms of conventional physics. He also felt that because none of the negative experiments used either the same sources or the same detectors that their evidential weight was lessened, a view not widely shared.[67]

Hughes et al. (1989) analyzed the Greenland result in terms of non-Newtonian gravity, although they recognized that the result could also be explained in terms of local density variations. They found an attractive force, with a strength between 2.4 % and 3.5 % of Newtonian gravity and a range between 225 m and 5.4 km. They argued that this result was consistent with that of Eckhardt et al. (1987b), although it was inconsistent with Stacey's measurements.

Shortly after the 1989 Moriond workshop the Greenland group ice cap measurement was published (Ander et al. 1989). As they had earlier, they reported a 3.87 mGal anomaly between the gravity values at depths of 213 and 1673 m, but their conclusion had changed (p. 985): "We cannot unambiguously attribute it to a breakdown of Newtonian gravity because we have shown it might be due to unexpected geological features below the ice."

---

[67] As seen from my earlier discussion, Fischbach and I have different views on the value of a variety of evidence.

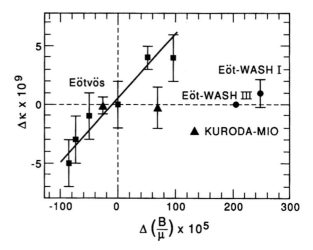

Fig. 2.28 The data of Kuroda and Mio (1989b) added to Fig. 2.20

Kuroda and Mio (1989b) also published the result they had presented at the Grossman conference. They reported no difference in gravitational acceleration for Al–C and Al–Cu pairs. If one includes their results, along with those of the Eöt-Wash group, and the original reanalysis of Fischbach et al., one finds that the striking effect presented by Fischbach has almost completely disappeared (Fig. 2.28).

At this time Keyser (1989) offered both a criticism and a possible explanation in terms of conventional physics of Thieberger's result, one of the major pieces of evidence for a Fifth Force. He noted that in a symmetric environment Thieberger had still found velocities approximately one third those obtained at the Palisades cliff. He also offered an explanation of those results in terms of convection. Thieberger (1989) replied that the velocities in the symmetric environment were randomly directed whereas they were in a single direction, perpendicular to the cliff, in the Palisades experiment. He also noted that the effect Keyser proposed should change sign as the coefficient of thermal expansion changed sign when the temperature of the water went from below 4 °C to above. This had, in fact, been checked during the original experiment and no drastic changes in either the magnitude or sign of the velocity had been observed. He was, however, quite aware of the novelty of his experimental apparatus and of the subsequent negative measurements (Thieberger 1989, p. 810): "The observed motion could indeed have been due to ordinary forces. Unanticipated spurious effects can easily appear when a new method is used for the first time to detect a weak signal. Neither the title nor the text of Thieberger (1987a) contains a claim to the discovery of a new force. [...] Even though the sites and the substances vary, effects of the magnitude expected from Thieberger (1987a) have not been observed. Therefore, although convection of the type proposed by Keyser does not seem to be the explanation, it now seems likely that some other spurious effect may have caused the motion observed at the Palisades cliff." I should emphasize here that Thieberger had not found such a spurious effect. He was responding to the inability of others to obtain results in agreement with his.

Bartlett and Tew (1989b) continued their work on the effect of local terrain. They published a calculation arguing that the Hilton mine data of Stacey et al. (1987b) could also be due to a failure to adequately include the terrain in their theoretical model. They noted that in 1984 Holding and Tuck had stated that "the topographic effects are insignificant," whereas their later results (Holding et al. 1986) claimed to include such corrections. Bartlett and Tew found the agreement of the two results unlikely if the terrain corrections had been made correctly, in view of the large terrain effect they had obtained. They had communicated their concerns to Stacey and were awaiting further developments. These were forthcoming. At the General Relativity and Gravitation Conference held 2–8 July 1989, Tuck (1989) reported that their group had incorporated a new and more extensive surface gravity survey in their calculation: "Preliminary analysis of these data indicates a regional bias that reduces the anomalous gravity gradient to two thirds of the value that we had previously reported (with a 50% uncertainty)."[68] With such a large uncertainty, these mineshaft results could certainly not be considered very positive evidence for the Fifth Force, if they provided any support at all.

Parker and Zumberge (1989), two members of the Greenland group, offered a general criticism. This provided more details than the earlier Parker paper (1988). They argued that they could explain the anomalies reported in both the tower experiment (Eckhardt et al. 1987b) and their own ice cap experiment using conventional physics and plausible local density variations. They had not been able to do this for the Australian mine result (Stacey et al. 1987b) because the original survey data were proprietary. They concluded that there was (Parker and Zumberge 1989, p. 31) "no compelling evidence for non-Newtonian long-range forces in the three most widely cited geophysical experiments; [...] and that the case for the failure of Newton's Law had not been established."

Toward the end of 1989, the Eöt-Wash group published their most stringent limits yet on the presence of a Fifth Force (Heckel et al. 1989). They found for Be–Al and Be–Cu test-body pairs $\Delta a = (1.5 \pm 2.3) \times 10^{-11}$ cm s^{-2} and $\Delta a = (0.9 \pm 1.7) \times 10^{-11}$ cm s^{-2}, respectively. They concluded that (p. 2707) "our null results are in strong disagreement with the positive effects observed by Thieberger, Boynton et al., Eckhardt et al., and Stacey et al." They did not, in fact, include either the Stacey or the Eckhardt results in their final figure (p. 2707), "because a previously unidentified systematic error has been discovered in these results." They cited private communications from both Stacey and Eckhardt as well as the published work of Bartlett and Tew (1989b).

The only other new experimental result presented in 1989 was by Muller et al. (1989). They measured gravity using six gravimeters in close proximity to a pumped-storage reservoir in which there were daily water variations of between 5 and 22 m (Muller et al. 1989, p. 2621): "The experiment's goal was a search

---

[68]To show that publication date may not reflect the real history, I note that the Bartlett and Tew paper was published on 15 July 1989, after the conference had been held. It was, however, submitted on 3 January 1989.

for deviations from Newton's gravitational law, but it can also be viewed as a measurement of the gravitational constant G for effective mass distances of 40–70 m. The deviation of G from the laboratory value was found to be $(0.25 \pm 0.40)\%$ and thus is not significantly different from zero."

2.5   The Force Is Not With Us

In a real sense the 1990 Moriond workshop (Fackler and Tran Tanh Van 1990) marked the last hurrah for the Fifth Force.[69] Stubbs (1990a), a member of the Eöt-Wash group, offered an introductory survey and summary of the evidential situation at the beginning of the conference. He noted that results on the composition-dependence of the Fifth Force included only two positive claims, those of Thieberger and those of Boynton et al. from the first Index experiment. He contrasted these with the 12 negative results already reported, and suggested that until the positive results were replicated, there was no need for other experimenters to work at either the Index or Palisades site, as some had suggested.[70] As far as the geophysical results were concerned, he reported that Stacey et al. had retracted their positive result and that Eckhardt had found a bias in his gravity survey and would be presenting new results at the workshop. He concluded (p. 185): "It seems clear at this point that the original 'Fifth Force' hypothesis of a coupling to baryon number is not consistent with experiment." The same could not yet be said for coupling to isospin.

Perhaps the most important experimental result presented at the workshop was that of Boynton (1990).[71] He reported on a continuation of the torsion pendulum experiment at Index (Index III) and concluded that (p. 207) "the Index III experiment sets the most stringent upper limits yet on the interaction strength for coupling to from $B - 2L$ (isospin) to $B - L$, and for an interaction range from 200 m to 10 km. It is also the first null result to conflict with the marginal detection reported for the Index I experiment for *all* (emphasis in original) relevant values of the composition and range parameters."[72] Boynton presented limits for ξ, for $\lambda = 100$ m, of $-4.3 \times 10^{-5} < \xi < 1.8 \times 10^{-4}$ and $3.2 \times 10^{-5} < \xi < 1.2 \times 10^{-4}$ for isospin and baryon coupling, respectively. These were more stringent than the range-independent limits

---

[69]Although, as discussed later, some new results would be published after the workshop, and some previously reported work as well as papers presented at the workshop would be published later, nothing really new was presented.

[70]Not everyone present at the workshop agreed. Although few, if any, scientists believe that there is anything pathological about the site of Thieberger's experiment, some nagging doubts remain. At the 1990 Moriond Workshop a petition was circulated asking Thieberger to repeat his experiment at the Palisades cliff, which had a considerable number of signatures. As of the moment, Theiberger has not repeated his experiment.

[71]Boynton listed his collaborators: S. Aronson, P. Ekstrom, D. Crosby, A. Eberhart, E. Lindahl, P. Peters, and M. Wensman.

[72]Boynton disagreed with the conclusion expressed in Heckel et al. (1989).

Fig. 2.29 Difference between measured and calculated values of *g* as a function of height (From Jekeli et al. 1990)

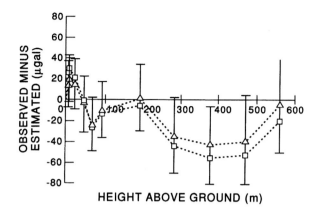

HEIGHT ABOVE GROUND (m)

on isospin coupling of $\xi < 4.6 \times 10^{-4}$ (Stubbs et al. 1989b), $\xi < 2.3 \times 10^{-4}$ (Cowsik et al. 1990), and $\xi < 1.8 \times 10^{-4}$ (Nelson et al. 1990, preprint). He still had no explanation of the Index I result, however.

The other important result was that of Eckhardt et al. (1989a) and Jekeli et al. (1990). The title of Jekeli et al. (1990) says it all *Tower Gravity Experiment: No Evidence for Non-Newtonian Gravity*. They reached the conclusion (Jekeli et al. 1990, p. 1204): "We have refined the analysis of that experiment [their 1988 result] by including detailed topographical information and conclude that, in fact, no such evidence [for non-Newtonian gravity] exists." This is clearly seen in Fig. 2.29.

Other groups also presented negative results. A new tower gravity experiment (Speake et al. 1990a,b) reported agreement between the measured value and the calculated Newtonian results at the top of the 300 m tower of $21 \pm 27\,\mu$Gal. The tower and its base are shown in Figs. 2.30 and 2.31. The Livermore tower group presented further analysis of their experiment, which continued to support Newton's law (Kammeraad et al. 1990), and Stubbs (1990b) presented the previously published Eöt-Wash results. Bizzeti et al. (1990) presented no new results from their float experiment, but did report on tests of the sensitivity of their experimental apparatus to gravity gradients. They did not change their conclusion that there was no Fifth force. They did, in addition, present an analysis of their data using Fischbach's suggested exponential force, and found limits similar to those they had found previously for the Yukawa model.

During 1990 two other groups Cowsik et al. (1990) and Kuroda and Mio (1990) presented more stringent limits resulting from improved experimental apparatuses. They continued to conclude that there was no evidence for a composition dependent force. Kuroda and Mio also presented limits on the exponential model.

By the end of 1990 virtually the only remaining experimental result supporting the Fifth Force was Thieberger's (1987a) float experiment. Although Boynton still had no explanation for the marginally detected (his words) results of Index I, these

Fig. 2.30 The 300 m tower in Erie, Colorado used by Speake et al. (1990a,b) for their gravity measurements (Courtesy of Jim Faller)

seem superseded by those of Index III. In view of the vast preponderance of evidence it seems fair to conclude: "The Fifth Force is not with us."[73]

---

[73]Everyone would certainly agree that this statement applies to the originally proposed Fifth Force—a force approximately 1 % that of normal gravity, with a range of the order of 100 m. Recall Fackler's comment at the 1990 workshop: "The Fifth Force is dead." As discussed below, however, experimental work is still continuing with the goal of setting more stringent limits on such a force or perhaps even of finding one.

Those physicists who worked on the Fifth Force always considered themselves as outsiders within the physics community and believed their work was not regarded as valuable. As evidence of this I present a proposed letter to Dear Abby that was circulated within the group, dated March

Fig. 2.31 Clive Speake and a LaCoste–Romberg gravimeter at the base of the Erie tower (Courtesy of Jim Faller)

7, 1990. The letter was written and circulated by Don Eckardt, who based it on something he had read earlier, although he doesn't recall what that was.

Dear Abby,
I have a problem. I have two brothers, one who is a scientist doing research on the Fifth Force and another who is sentenced to death in the electric chair for a series of homosexual rapes and murders. My mother died from insanity when I was three years old. She had syphilis and I think that I caught it from her. My two sisters are prostitutes, and my father is now selling pornography and kinky sexual paraphernalia following his bust for retailing narcotics. Recently I met a young girl who had just been released from an institution for the criminally insane where she had served time for smothering her illegitimate child. I love

2.6 Some Reasons Why

In the introduction to this study I suggested several reasons why a scientist might choose to further investigate, or to pursue, a hypothesis. These included the interest and importance of the hypothesis; its plausibility, based on existing evidence, on its resemblance to other successful theories, or on its mathematical properties; the fact that it fit in with an ongoing research program; and its ease of test, in which I include the conceptual simplicity of the test, which differs from the technical experimental details of the test, which might be quite complex; and whether or not the experiment can be performed with either existing apparatus or with small modifications of it, or with a relatively modest investment in a new apparatus.

Because the search for the Fifth Force has taken place, almost entirely, within the last six years (as discussed below, it is, in a sense, still continuing) all of the major participants are available for discussion. Although such discussions can provide only anecdotal evidence (I have not made a statistical survey of a large number of the scientists involved), they are, I believe, a reasonable way to examine the issue of pursuit. After all, who knows better than the participants themselves why they worked on something? I have spoken with several of the scientists involved in the investigation of the Fifth Force and discussed their initial involvement with them. These were Eric Adelberger, David Bartlett, Paul Boynton, Donald Eckhardt, James Faller, and Riley Newman.[74] With the exception of Bartlett, an experimentalist whose contribution here was primarily theoretical or calculational, they are all experimentalists. Theorists had been working on modifications of the 1961 Brans–Dicke theory of gravitation since the early 1970s, although, as discussed earlier, the publication of the Fifth Force hypothesis certainly stimulated new theoretical work and also gave added impetus to ongoing programs.

In three cases, those of Eckhardt, Faller, and Newman, the experiments done fitted in with an ongoing research program. Eckhardt, a scientist in the Air Force Geophysical Laboratory, had been planning balloon measurements of gravity in order to investigate whether or not the lack of detailed and precise knowledge of surface gravity might account for missile accuracy problems. One of the important factors in such a measurement is knowledge of the exact position of the balloon and Eckhardt had planned to use the Global Positioning System to determine it. Unfortunately, the system was not yet operating so the planned measurements had to be delayed. When the original Fifth Force paper appeared it gave added

---

this girl very much and I want to marry her. She loves me too, even though I have AIDS. My problem is this: should I tell her about my brother who is working on the Fifth Force?

Yours truly,
Bewildered

[74]I have already mentioned extensive discussions with Ephraim Fischbach and Sam Aronson. I also had an opportunity to speak with other participants at the Moriond Workshops of 1989 and 1990.

impetus to such measurements and encouraged Eckhardt to perform his tower gravity experiment.

Faller had been working on gravity experiments since his days as a graduate student at Princeton in the early 1960s. (His dissertation was *An Absolute Inter-ferometric Determination of Gravity* [Princeton University, 1963].) At the time the Fifth Force paper was published he was working on an experiment to measure g, the acceleration due to gravity at the surface of the Earth. The experiment involved dropping a weight in a specially constructed chamber. He was also constructing a second chamber for use by another group. He realized that with relatively modest modifications he could use both chambers for a Galileo-type test of the composition dependence of the gravitational force (or the Fifth Force) and proceeded to do so. He reports that these relatively modest modifications took six months to complete.

Newman had been working on the distance dependence of the inverse square law of gravity since about 1980, motivated by Long's work, discussed earlier. He was both a participant in and a coauthor of two experiments that had set very stringent limits on the deviations from the inverse square law at very short distances (of the order of a few centimeters) discussed earlier (Spero et al. 1980; Hoskins et al. 1985). Even prior to the publication of the Fifth Force paper he had made a proposal to the National Science Foundation for an experiment to investigate the possible composition dependence of the gravitational force.

For the other scientists interviewed, the investigations of this "intriguing possibil-ity," a phrase used by both Adelberger and Boynton, involved changing their area of research. All of the researchers remarked that the idea of testing a fundamental law of physics with a table-top experiment, or with a comparatively inexpensive and conceptually simple apparatus, was an important part of their motivation. Several of these investigators had worked previously on tests of other fundamental laws. Bartlett, for example, had worked on tests of time reversal violation, a fundamental symmetry in nature, and on the distance dependence of the inverse square law of electrostatic force (Coulomb's Law). Newman had, in addition to his earlier work on gravity, also investigated whether or not there was a spatial asymmetry in beta decay. Adelberger noted that he had discussed the possibility of a Fifth Force test with Heckel, another member of the Eöt-Wash group, and that they had been able to come up with both a relatively simple idea for a workable apparatus along with possible background and systematic effects within an hour. They had both, in their respective work in nuclear and atomic physics, previously worked on the measurement of small effects.[75]

For all of these experimenters the original Fifth Force paper acted as the imme-diate cause of their subsequent work. Adelberger and Boynton, faculty members at the University of Washington, also stated that their motivation had been enhanced by a seminar given at Washington by Fischbach during the spring of 1986. One may speculate that the attention given to the Fifth Force in the popular press, along with the large number of talks given on the subject by both Fischbach and Aronson

---

[75]Adelberger had been awarded the Bonner Prize of the American Physical Society for his work in nuclear physics.

(recall that in the six months following the publication Fischbach gave 16 talks and Aronson gave 14) also helped to stimulate subsequent work. It certainly gave the hypothesis a wide audience.

Bartlett entered the field later and his work was motivated by the anomalous results reported by the Greenland group (their earliest report), by Eckhardt, and by Stacey. He was quite skeptical of all of these results and thought that they must be due to some kind of local variation. He remarked that he thought of terrain as the most probable cause within about two weeks. He also noted that this was, fortunately, rather easy to investigate because of the availability of topographical maps for both the Australian (Stacey) and North Carolina (Eckhardt) sites.

An interesting point here involving the context of pursuit is that investigating a hypothesis does not necessarily require a belief in its truth. It is fair to say that the physicists I spoke with were quite skeptical about the existence of a Fifth Force.[76] Their attitudes ranged from Eckhardt's view that Fischbach et al. (1986a) were wrong and that he was going to demonstrate it with his tower experiment, to Newman's belief that the hypothesis had a 20–30 % chance of being correct, not an overwhelmingly positive view. Adelberger remarked on the difficulty of both analyzing the data and of finding systematic effects in current experimental work and expressed doubts that they could be done well for experiments performed 75 years earlier, although he found the results of the reanalysis very interesting. One should recall here that Eckhardt, despite his expressed skepticism, reported a positive Fifth Force result, at least in his early work, and that Newman, who was more positive, reported an experiment that found no such effect and set some of the more stringent limits on such a force. Scientists do not always find what they are looking for.[77]

This is not to say that evidence was totally unimportant in the decision to pursue a research program, but rather that a hypothesis can act as a stimulus for further work even if one were skeptical of both the hypothesis and the evidence supporting it. It seems clear that the fact that the original paper contained the reanalyzed Eötvös data made the hypothesis of the Fifth Force more plausible and led to the subsequent work. Recall that similar theoretical work, by Fujii and others, had been going on since the early 1970s without stimulating the large amount of work that followed publication of the Fifth Force hypothesis.

---

[76]It might be suggested that given the subsequent demise of the Fifth Force that the participants may now report views that differ from those they held at the time. That they may now claim never to have believed in the Fifth Force even if they had originally held more positive views. There is always a danger that people may recount a story in the manner that makes them look best. I don't believe that this is a problem in this case because the views expressed are consistent with those given by the participants at the Moriond workshops, before the issue was resolved.

[77]In fact, the earliest run of the Eöt-Wash experiment actually gave a positive result. The series of negative results reported by this group were among the strongest arguments against the existence of a Fifth Force. Adelberger, who was informed of this result by telephone while he was visiting at the University of Wisconsin, was "tremendously surprised". The experimenters then rotated the mirror in the apparatus by 90° (see Fig. 2.6) and found that the signal did not change phase as it was expected to if it were a valid signal. A systematic source of error was subsequently found and corrected.

2.7 Epilogue

Although virtually everyone agrees that the Fifth Force, at least in its originally suggested form of a composition dependent force with a strength of approximately 1 % that of gravity and with a range of the order of 100 m, does not exist, experimental work is still continuing. Four of the experimenters mentioned in the previous section are currently continuing their experiments. In part, this is because their experience in working with the apparatus has allowed them to learn about backgrounds and sources of systematic error and has allowed them to design and construct experiments with greater precision and accuracy. One might reasonably call this "instrumental loyalty" and "recycling of expertise". These experiments will set more stringent limits on the presence of a violation of the law of gravity or measure a weaker force, if it exists. It is, as Adelberger has already stated, a rather cost effective method for searching for new physics.

The Eöt-Wash group is constructing two torsion pendulums. One will be used in the continuation of their previous work using a hillside source, while the other, which will be stationary and use a moving laboratory mass, will measure the distance dependence of the gravitational force down to a limit of about 2 cm. Boynton has also constructed two new instruments. One is portable and will be used at both Mount Index, the site of his earlier work, and at the Palisades cliff, the site of Thieberger's experiment. A second instrument will use a laboratory source consisting of 1 ton of lead, and is designed to investigate the distance dependence down to fractions of a centimeter. In both cases, Eöt-Wash and Boynton, the experiments will have greater precision than previously. Boynton remarked, however, that he did not expect these new experiments to give the source of previous errors or to explain the earlier positive results.

Newman is also continuing his experiment with the goal of greater precision. Eckhardt is currently working on a tower experiment at a site in Mississippi, at which the terrain calculation should be more straightforward than it was at his North Carolina site, and hopes to set more stringent limits on any possible anomaly than any previous tower experiment.

Interestingly, Bartlett, whose discussion of the effects of terrain cast doubt on claimed positive Fifth Force results and led to their correction, is currently working on terrain with the Eöt-Wash group. It seems only fair to him that he now work on an experiment that gave negative results. Fischbach and Talmadge are assisting on Eckhardt's tower experiment (see Fig. 2.32).

The most recent measurement of G, the gravitational constant, on a macroscopic scale of 500 m, done by Zumberge et al. (1991), gave a value of $G = (6.677 \pm 0.013) \times 10^{-11}\, \mathrm{m^3\, s^{-2}\, kg^{-1}}$ which was consistent, within the experimental uncertainty, with the best laboratory value of 6.6726×10^{-11}. This was a carefully done experiment involving gravity measurements along a vertical ocean profile by a submersible, along submerged horizontal planes by a submarine, along the ocean bottom by a remotely operated gravity meter, and along the ocean surface by a shipboard gravity meter. The terrain was chosen to minimize corrections for

Fig. 2.32 Carrick Talmadge positioning an antenna for the Global Positioning Satellite as part of a current tower gravity experiment, which also includes Eckhardt and Fischbach. Note the flatness of the terrain (Courtesy of Ephraim Fischbach)

such effects and the seawater density was measured as a function of depth, and the effects of local density variations estimated (p. 3054): "Roughly speaking, this result constrains the magnitude of the coupling constant of a single Yukawa modification to Newtonian gravity to be less than 0.002 for scale lengths in the range from 1 m to a few km."

A recent review of the subject Adelberger et al. (1991) concludes (see Table 2.5) (Adelberger et al. 1991, p. 306):

> Considerable experimental progress has occurred in the four years since 1986 when Fischbach et al. proposed a "fifth force". New experimental techniques have been introduced, and sensitivities have increased dramatically. The situation regarding inverse-square law

Table 2.5 Modern tests of the universality of free fall (From Adelberger et al. 1991)

Det $\Delta(B/\mu)(10^{-4})$	Source[a] (B/μ)	$\alpha\Delta(B/\mu)_d(B/\mu)_s^c$				Reference
		$\lambda = 1\,\mathrm{m}$	$\lambda = 30\,\mathrm{m}$	$\lambda = 1000\,\mathrm{m}$	$\lambda = \infty$	
7.0	1.00		$(0.4 \pm 1.4)10^{-4}$	$(1.1 \pm 4.3)10^{-6}$	$(1.3 \pm 5.1)10^{-10}$	Niebauer et al. (1987)
17.6	1.00		$(1.2 \pm 3.5)10^{-4}$	$(0.4 \pm 1.1)10^{-5}$	$(0.4 \pm 1.3)10^{-9}$	Kuroda et al. (1990)
4.3	1.00		$(0.4 \pm 2.2)10^{-4}$	$(1.1 \pm 6.6)10^{-6}$	$(1.3 \pm 8.0)10^{-10}$	Kuroda et al. (1990)
6.4	1.00		$(-0.5 \pm 3.9)10^{-4}$	$(-0.2 \pm 1.2)10^{-5}$	$(-0.2 \pm 1.4)10^{-9}$	Kuroda et al. (1990)
5.14	0.999				$(-1.3 \pm 1.5)10^{-11}$	Roll et al. (1964)
5.01	0.994				$(3.0 \pm 4.5)10^{-13}$	Braginskii et al. (1972)
20.36	1.0006	$(0.6 \pm 1.4)10^{-6}$	$(1.5 \pm 0.4)10^{-6}$	$(1.2 \pm 0.3)10^{-7}$	—[d]	Boynton et al. (1987)
24.69	1.0005		$(1.4 \pm 2.9)10^{-8}$	$(1.4 \pm 4.2)10^{-9}$	$(-0.2 \pm 1.0)10^{-11}$	Heckel et al. (1989)
20.36	1.0005	$(-0.9 \pm 1.7)10^{-6}$	$(-2.1 \pm 3.6)10^{-8}$	$(-5.1 \pm 5.1)10^{-9}$	$(-0.5 \pm 1.3)10^{-11}$	Heckel et al. (1989)
22.35	1.00		$(-3.0 \pm 5.2)10^{-6}$	$(-1.1 \pm 2.0)10^{-7}$	$(1.8 \pm 12.9)10^{-9}$	Fitch et al. (1988)
10.01	0.00098[b]	$(-7.2 \pm 7.6)10^{-4}$	$(-7.2 \pm 7.6)10^{-4}$	$(-7.2 \pm 7.6)10^{-4}$	$(-7.2 \pm 7.6)10^{-4}$	Speake et al. (1988)
1.19	0.00098[b]	$(-3.2 \pm 3.7)10^{-4}$	$(-3.2 \pm 3.7)10^{-4}$	$(-3.2 \pm 3.7)10^{-4}$	$(-3.2 \pm 3.7)10^{-4}$	Speake et al. (1988)
10.01	0.9994		$(-0.9 \pm 1.9)10^{-5}$	$(-0.7 \pm 1.4)10^{-5}$	$(-0.7 \pm 1.4)10^{-5}$	Bennett (1989a)
20.36	1.0001	$(0.5 \pm 4.1)10^{-6}$	$(0.4 \pm 3.9)10^{-6}$	$(0.4 \pm 3.9)10^{-6}$	$(0.4 \pm 3.9)10^{-6}$	Stubbs et al. (1989b)
10.01	0.00098[b]		$(-1.4 \pm 0.9)10^{-6}$	$(-1.4 \pm 0.9)10^{-6}$	$(-1.4 \pm 0.9)10^{-6}$	Cowsik et al. (1990)
9.52	1.0002	$(1.1 \pm 1.2)10^{-6}$	$(1.1 \pm 1.2)10^{-6}$	$(1.1 \pm 1.2)10^{-6}$	$(1.1 \pm 1.2)10^{-6}$	Nelson et al. (1990)
17.04	1.0006		$(-6.8 \pm 2.2)10^{-5}$	$(-5.1 \pm 1.7)10^{-6}$	—[d]	Thieberger (1987b)
3.73	1.0006			$(0.0 \pm 1.1)10^{-7}$	$(0.0 \pm 1.4)10^{-9}$	Bizzeti et al. (1989b)

[a]Charges of terrestrial sources depend slightly on λ, but this has little practical significance

[b]Because of the geometry of the source these are $\Delta(B/\mu)$ values

[c]Uncertainties are $\pm 1\sigma$ errors

[d]A positive effect was observed in a direction not consistent with $\lambda = \infty$

tests is now clear. No violations are observed in astronomical or laboratory experiments. The claims of $1/r^2$ violations in geophysical tests that probed $g(z)$ in boreholes and on towers have all been retracted, and replaced by improved upper limits on any violation of Gauss' Law. The earlier erroneous claims are now understood to have been due to inadequate accounting for the local terrain.

The experimental situation in tests of the universality of free fall [composition dependence], summarized in Table 2.5, has also been greatly improved. With two exceptions, experiments show no evidence for a new macroscopic interaction. The most sensitive (in terms of differential acceleration resolution) results in each category—von Eötvös experiments with Earth and laboratory sources, Galileo experiments, and floating ball experiments—give null results.[78]

The Fifth Force may indeed be dead, but work continues. Experiments do seem to generate a life of their own and I will discuss this in the next section.

References

Adelberger, E.G.: High-sensitivity hillside results from the Eöt-Wash experiment. In: Fackler, O., Tran Thanh Van, J. (eds.), pp. 485–499. Editions Frontières, Gif sur Yvette (1989)

Adelberger, E.G., et al.: New constraints on composition-dependent interactions weaker than gravity. Phys. Rev. Lett. **59**, 849–852 (1987)

Adelberger, E.G., et al.: Constraints on composition-dependent interactions from the Eöt-Wash experiment. In: Fackler, O., Tran Thanh Van, J. (eds.) Fifth Force Neutrino Physics: Eighth Moriond Workshop, pp. 445–456. Editions Frontières, Gif sur Yvette (1988)

Adelberger, E.G., et al.: Searches for new macroscopic forces. Ann. Rev. Nucl. Part. Sci. **41**, 269–320 (1991)

Airy, G.B.: Account of pendulum experiments undertaken in the Harton Colliery, for the purpose of determining the mean density of the earth. Philos. Trans. R. Soc. Lond. **146**, 297–355 (1856)

Ander, M.E., et al.: Test of Newton's inverse-square law in the Greenland Ice Cap. Phys. Rev. Lett. **62**, 985–988 (1989)

Aronson, S.H., et al.: Experimental signals for hyperphotons. Phys. Rev. Lett. **56**, 1342–1345 (1986)

Avron, Y., Livio, M.: Considerations regarding a space-shuttle measurement of the gravitational constant. Astrophys. J. **304**, L61–64 (1986)

---

[78]The authors go on to discuss the anomalous results (Adelberger et al. 1991, p. 306):

> Although reports of positive effects by Thieberger and by Boynton et al. have not been retracted, these authors themselves do not claim evidence for new physics. However, because no two experiments are alike, there is always the possibility that the positive effects occurred because some special feature of the detectors or sources used by Thieberger or Boynton et al. allow them to see new physics that was not detected in other experiments.

They then discuss the possibility of more complex Fifth Force scenarios, including coupling to quantities other than baryon number, and multicomponent Yukawa potentials, or non-Yukawa potentials. They conclude [p. 306]: "We prefer to adopt 'Occam's razor' and, until the positive results have been reproduced, assume they are due to as yet unidentified systematic errors in very difficult experiments." This discussion does not, I believe, alter the conclusion that there is no Fifth Force.

Barr, S.M., Mohapatra, R.N.: Range of Feeble forces from higher dimensions. Phys. Rev. Lett. **57**, 3129–3132 (1986)

Bars, I., Visser, M.: Feeble intermediate-range forces from higher dimensions. Phys. Rev. Lett. **57**, 25–28 (1986)

Bars, I., Visser, M.: Feeble forces and gravity. Gen. Rel. Gravitat. **19**, 219–223 (1987)

Bartlett, D.F., Tew, W.L.: The Fifth Force: Terrain and pseudoterrain. In: Fackler, O., Tran Thanh Van, J. (eds.), pp. 543–548. Editions Frontières, Gif sur Yvette (1989a)

Bartlett, D.F., Tew, W.L.: Possible effect of the local Terrain on the Australian fifth-force measurement. Phys. Rev. D **40**, 673–675 (1989b)

Bartlett, D.F., Tew, W.L.: Terrain and geology near the WTVD tower in North Caroline: implications for non-Newtonian gravity. J. Geophys. Res. **95**(17), 363–369 (1990)

Bennett, W.R.: Modulated-source Eötvös experiment at little goose lock. Phys. Rev. Lett. **62**, 365–368 (1989a)

Bennett, W.R.: Bennett replies. Phys. Rev. Lett. **63**, 810 (1989b)

Bertolami, O.: Testing the Baryon number of hypercharge interaction with a neutron interferometric device. Mod. Phys. Lett. A **1**, 383–388 (1986)

Bizzeti, P.G.: Significance of the Eötvös method for the investigation of intermediate range forces. Nuovo Cim. B **94**, 80–86 (1986)

Bizzeti, P.G.: Forces on a floating body: another way to search for long-range, B-dependent interactions. In: Fackler, O., Tran Thanh Van, J. (eds.), pp. 591–598. Editions Frontières, Gif sur Yvette (1987)

Bizzeti, P.G., et al.: New search for the Fifth Force with the floating-body method: status of the Vallambrosa experiment. In: Fackler, O., Tran Thanh Van, J. (eds.), pp. 501–513. Editions Frontières, Gif sur Yvette (1988)

Bizzeti, P.G., et al.: Search for a composition dependent Fifth Force: results of the Vallombrosa experiment. In: Fackler, O., Tran Thanh Van, J. (eds.), pp. 511–524. Editions Frontières, Gif sur Yvette (1989a)

Bizzeti, P.G., et al.: Search for a composition-dependent Fifth Force. Phys. Rev. Lett. **62**, 2901–2904 (1989b)

Bizzeti, P.G., et al.: Recent tests of the Vallambrosa experiment. In: Fackler, O., Tran Thanh Van, J. (eds.) New and Exotic Phenomena '90: Tenth Moriond Workshop, pp. 263–268. Editions Frontières, Gif sur Yvette (1990)

Blair, D.G., Buckingham, M.J. (eds.): Proceedings of the Fifth Marcel Grossman on General Relativity. World Scientific, Singapore (1989)

Bouchiat, C., Iliopoulos, J.: On the possible existence of a light vector Meson coupled to the hypercharge current. Phys. Lett. B **169**, 447–449 (1986)

Boynton, P.E., et al.: Search for an intermediate-range composition-dependent force. Phys. Rev. Lett. **59**, 1385–1389 (1987)

Boynton, P., Peters, P.: Torsion pendulums, fluid flows and the coriolis force. In: Fackler, O., Tran Thanh Van, J. (eds.), pp. 501–510. Editions Frontières, Gif sur Yvette (1989)

Boynton, P.: New limits on the detection of a composition-dependent macroscopic force. In: Fackler, O., Tran Thanh Van, J. (eds.) New and Exotic Phenomena '90: Tenth Moriond Workshop, pp. 207–224. Editions Frontières, Gif sur Yvette (1990)

Braginskii, V.B., Panov, V.I.: Verification of the equivalence of inertial and gravitational mass. JETP **34**, 463–466 (1972)

Cavasinni, V., et al.: Galileo's experiment on free-falling bodies using modern optical techniques. Phys. Lett. A **116**, 157–161 (1986)

Cho, Y.M.: Internal gravity. Phys. Rev. D **35**, 2628–2631 (1987)

Chu, S.Y., Dicke, R.H.: New force or thermal gradient in the Eötvös experiment? Phys. Rev. Lett. **57**, 1823–1824 (1986)

Colella, R., Overhauser, A.W., Werner, S.A.: Observation of gravitationally induced quantum interference. Phys. Rev. Lett. **34**, 1472–1474 (1975)

Cowsik, R., et al.: Limit on the strength of intermediate-range forces coupling to isospin. Phys. Rev. Lett. **61**, 2179–2181 (1988)

Cowsik, R., et al.: Strength of intermediate-range forces coupling to isospin. Phys. Rev. Lett. **64**, 336–339 (1990)

De Rujula, A.: Are there more than four? Nature **323**, 760–761 (1986a)

De Rujula, A.: On weaker forces than gravity. Phys. Lett. B **180**, 213–220 (1986b)

D'Olivo, J.C., Ryan, M.P.: A Newtonian cosmology based on a Yukawa-type potential. Class. Quantum Gravity **4**, 113–116 (1987)

Eckhardt, D.H.: Comment on 'reanalysis of the Eötvös experiment.' Phys. Rev. Lett. **57**, 2868 (1986)

Eckhardt, D.H., et al.: Results of a tower gravity experiment. In: Fackler, O., Tran Thanh Van, J. (eds.), pp. 577–583. Editions Frontières, Gif sur Yvette (1987a)

Eckhardt, D.H., et al.: Tower gravity experiment: evidence for non-Newtonian gravity. Phys. Rev. Lett. **60**, 2567–2570 (1987b)

Eckhardt, D.H., et al.: Evidence for non-Newtonian gravity: status of the AFGL experiment January 1989. In: Fackler, O., Tran Thanh Van, J. (eds.), pp. 525–527. Editions Frontières, Gif sur Yvette (1989a)

Eckhardt, D.H., et al.: Detection of non-Newtonian gravity: the AFGL tower gravity experiment. In: Blair, D.G., Buckingham, M.J. (eds.) Proceedings of the Fifth Marcel Grossman on General Relativity, pp. 1565–1568. World Scientific, Singapore (1989b)

Elizalde, E.: About the Eötvös experiment and the hypercharge theory. Phys. Lett. A **116**, 162–166 (1986)

Fackler, O., Tran Thanh Van, J. (eds.): New and Exotic Phenomena: Seventh Moriond Workshop. Editions Frontières, Gif sur Yvette (1987)

Fackler, O., Tran Thanh Van, J. (eds.): Fifth Force Neutrino Physics: Eighth Moriond Workshop. Editions Frontières, Gif sur Yvette (1988)

Fackler, O., Tran Thanh Van, J. (eds.): Tests of Fundamental Laws in Physics: Ninth Moriond Workshop. Editions Frontières, Gif sur Yvette (1989)

Fackler, O., Tran Thanh Van, J. (eds.): New and Exotic Phenomena '90: Tenth Moriond Workshop. Editions Frontières, Gif sur Yvette (1990)

Fairbank, W.M.: Summary talk on the Fifth Force papers. In: Fackler, O., Tran Thanh Van, J. (eds.), pp. 629–644. Editions Frontières, Gif sur Yvette (1988)

Fayet, P.: A new long-range force? Phys. Lett. B **171**, 261–266 (1986)

Fischbach, E.: Multicomponent models of the Fifth Force. In: Fackler, O., Tran Thanh Van, J. (eds.), pp. 541–556. Editions Frontières, Gif sur Yvette (1987)

Fischbach, E., et al.: Reanalysis of the Eötvös experiment. Phys. Rev. Lett. **56**, 3–6 (1986a)

Fischbach, E., et al.: Response to Thodberg. Phys. Rev. Lett. **56**, 2424 (1986b)

Fischbach, E., et al.: Response to Keyser, Niebauer, and Faller. Phys. Rev. Lett. **56**, 2426 (1986c)

Fischbach, E., et al.: Response to Eckhardt. Phys. Rev. Lett. **57**, 2869 (1986d)

Fischbach, E., et al.: Alternative explanations of the Eötvös results. Phys. Rev. Lett. **57**, 1959 (1986e)

Fischbach, E., et al.: A new force in nature? In: Geesaman, D. (ed.) AIP Conference Proceedings 150, Interactions Between Particle and Nuclear Physics, Lake Louise, pp. 1102–1118. American Institute of Physics, New York (1986f)

Fischbach, E., et al.: The Fifth Force. In: Loken, S.C. (ed.) Proceedings of the Twenty-Third International Conference on High Energy Physics, Berkeley, 16–23 July, pp. 1021–1031. World Scientific, Singapore (1986g)

Fischbach, E., et al.: Long-range forces and the Eötvös experiment. Ann. Phys. **182**, 1–89 (1988)

Fischbach, E., Talmadge, C., Sudarsky, D.: Alternative models of the Fifth Force. In: Fackler, O., Tran Thanh Van, J. (eds.), pp. 445–458. Editions Frontières, Gif sur Yvette (1989)

Fitch, V.L., Isaila, M.V., Palmer, M.A.: Limits on the existence of a material-dependent intermediate-range force. Phys. Rev. Lett. **60**, 1801–1804 (1988)

Franklin, A.: The Neglect of Experiment. Cambridge University Press, Cambridge (1986)

Franklin, A., Howson, C.: Why do scientists prefer to vary their experiments? Stud. Hist. Philos. Sci. **15**, 51–62 (1984)

Fujii, Y.: Theoretical models for possible nonzero effect in the Eötvös experiment. Prog. Theor. Phys. **76**, 325–328 (1986)

Fujii, Y.: Scalar–vector model of the Fifth Force. In: Fackler, O., Tran Thanh Van, J. (eds.), pp. 395–400. Editions Frontières, Gif sur Yvette (1988)

Gibbons, G.W., Whiting, B.F.: Newtonian gravity measurements impose constraints on unification theories. Nature **291**, 636–638 (1981)

Gilliland, R.L., Dappen, W.: Hypercharge, solar structure, and stellar evolution. Astrophys. J. **313**, 429–431 (1987)

Glass, E.N., Szamosi, G.: Intermediate-range forces and stellar structure. Phys. Rev. D **35**, 1205–1208 (1987)

Goldman, T., Hughes, R.J., Nieto, M.M.: Experimental evidence for quantum gravity? Phys. Lett. B **171**, 217–222 (1986)

Goldman, T., Hughes, R.J., Nieto, M.M.: Quantum gravity and the gravitiational acceleration of antimatter vs. matter. In: Fackler, O., Tran Thanh Van, J. (eds.), pp. 613–619. Editions Frontières, Gif sur Yvette (1987a)

Goldman, T., Hughes, R.J., Nieto, M.M.: Gravitational acceleration of antiprotons and of positrons. Phys. Rev. D **36**, 1254–1256 (1987b)

Grifols, J.A., Masso, E.: Constraints on finite-range baryonic and leptonic forces from Stellar evolution. Phys. Lett. B **173**, 237–240 (1986)

Grossman, N., et al.: Measurement of the lifetime of K_S^0 mesons in the momentum range 100–350 GeV/c. Phys. Rev. Lett. **59**, 18–21 (1987)

Hayashi, K.: A comment on space experiments of the 5th force. Europhys. Lett. **4**, 959–962 (1987)

Hayashi, K., Shirafuji, T.: Interpretation of geophysical and Eötvös anomalies. Prog. Theor. Phys. **76**, 563–566 (1986)

Hayashi, K., Shirafuji, T.: Is Thieberger's result inconsistent with Stubbs et al.'s one? Prog. Theor. Phys. **78**, 189–193 (1987a)

Hayashi, K., Shirafuji, T.: Constraints for free fall experiments undertaken on a substance dependent force. Prog. Theor. Phys. **78**, 22–26 (1987b)

Heckel, B.R., et al.: Experimental bounds on interactions mediated by ultralow-mass Bosons. Phys. Rev. Lett. **63**, 2705–2708 (1989)

Hills, J.G.: Space measurements of the gravitational constant using an artificial binary. Astron. J. **92**, 986–988 (1986)

Hinze, W.J., et al.: Gravimeter survey in the Michigan Basin deep borehole. J. Geophys. Res. (Solid Earth) **83**, 5864–5868 (1978)

Holding, S.C., Tuck, G.J.: A new mine determination of the Newtonian gravitational constant. Nature **307**, 714–716 (1984)

Holding, S.C., Stacey, F.D., Tuck, G.J.: Gravity in mines—an investigation of Newton's law. Phys. Rev. D **33**, 3487–3494 (1986)

Hoskins, J.K., et al.: Experimental tests of the gravitational inverse-square law for mass separations from 2 to 105 cm. Phys. Rev. D **32**, 3084–3095 (1985)

Hsui, A.T.: Borehole measurement of the Newtonian gravitational constant. Science **237**, 881–883 (1987)

Hughes, R.J., Goldman, T., Nieto, M.M.: Quantum gravity and new forces. In: Fackler, O., Tran Thanh Van, J. (eds.), pp. 603–607. Editions Frontières, Gif sur Yvette (1988)

Hughes, R.J., Goldman, T., Nieto, M.M.: Non-Newtonian gravity and the Greenland ice-sheet experiment. In: Fackler, O., Tran Thanh Van, J. (eds.), pp. 549–554. Editions Frontières, Gif sur Yvette (1989)

Hussain, A., et al.: Geophys. Prospect. **29**, 407 (1981)

Jekeli, C., Eckhardt, D.H., Romaides, A.J.: Tower gravity experiment: no evidence for non-Newtonian gravity. Phys. Rev. Lett. **64**, 1204–1206 (1990)

Kammeraad, J., et al.: New results from Nevada: a test of Newton's law using the BREN tower and a high density ground gravity survey. In: Fackler, O., Tran Thanh Van, J. (eds.) New and Exotic Phenomena '90: Tenth Moriond Workshop, pp. 245–254. Editions Frontières, Gif sur Yvette (1990)

Kasameyer, P., et al.: A test of Newton's law of gravity using the BREN tower, Nevada. In: Fackler, O., Tran Thanh Van, J. (eds.), pp. 529–542. Editions Frontières, Gif sur Yvette (1989)

Keyser, P.T.: Forces on the Thieberger accelerometer. Phys. Rev. Lett. **62**, 2332 (1989)

Keyser, P.T., Niebauer, T., Faller, J.E.: Comment on 'reanalysis of the Eötvös experiment'. Phys. Rev. Lett. **56**, 2425 (1986)

Kim, Y.E.: The local Baryon Gauge invariance and the Eötvös experiment. Phys. Lett. B **177**, 255–259 (1986)

Kim, Y.E.: New force or thermal convection in the differential-accelerometer experiment? Phys. Lett. B **192**, 236–238 (1987)

Kreuzer, L.B.: Experimental measurement of the equivalence of active and passive gravitational mass. Phys. Rev. **169**, 1007–1012 (1968)

Kuhn, J.R., Kruglyak, L.: Non-Newtonian forces and the invisible mass problem. Astrophys. J. **313**, 1–12 (1987)

Kuroda, K.: New force: the test of the equivalence principle. In: Fackler, O., Tran Thanh Van, J. (eds.), pp. 607–612. Editions Frontières, Gif sur Yvette (1987)

Kuroda, K., Mio, N.: Galilean test for composition-dependent force. In: Blair, D.G., Buckingham, M.J. (eds.) Proceedings of the Fifth Marcel Grossman on General Relativity, pp. 1569–1572. World Scientific, Singapore (1989a)

Kuroda, K., Mio, N.: Test of a composition-dependent force by a free-fall interferometer. Phys. Rev. Lett. **62**, 1941–1944 (1989b)

Kuroda, K., Mio, N.: Limits on a possible composition-dependent force by a Galilean experiment. Phys. Rev. D **42**, 3903–3907 (1990)

Li, M., Ruffini, R.: Radiation of new particles of the fifth interaction. Phys. Lett. A **116**, 20–24 (1986)

Long, D.R.: Comment on 'modulated-source Eötvös experiment at little goose lock'. Phys. Rev. Lett. **63**, 809 (1989)

Lusignoli, M., Pugliese, A.: Hyperphotons and K-meson decays. Phys. Lett. B **171**, 468–470 (1986)

Maddox, J.: Newtonain gravitation corrected. Nature **319**, 173 (1986a)

Massa, F.: Relevance of an intermediate-range force for Neutron–Antineutron oscillation experiments. Europhys. Lett. **2**, 87–90 (1986)

McCulloh, T.H.: A confirmation by gravity measurements of an underground density profile based on core densities. Geophysics **30**, 1108–1132 (1965)

Milgrom, M.: On the use of Eötvös-type experiments to detect medium-range forces. Nucl. Phys. B **277**, 509–512 (1986)

Moffat, J.: Nonsymmetric gravitation theory: a possible new force in nature. In: Tran Thanh Van, J. (ed.) Progress in Electroweak Interactions, Proceedings of the Lepton Section of the Twenty-First Rencontre de Moriond, Workshop, pp. 623–635. Editions Frontières, Gif sur Yvette (1986)

Moore, G.I., et al.: Determination of the gravitational constant at an effective mass separation of 22 m. Phys. Rev. D **38**, 1023–1029 (1988)

Muller, G., et al.: Determination of the gravitational constant by an experiment at a pumped-storage reservoir. Phys. Rev. Lett. **63**, 2621–2624 (1989)

Nelson, P.G., Graham, D.M., Newman, R.D.: Search for an intermediate-range composition-dependent force coupling to N–Z. Phys. Rev. D **42**, 963–976 (1990)

Neufeld, D.A.: Upper limit on any intermediate-range force associated with Baryon number. Phys. Rev. Lett. **56**, 2344–2346 (1986)

Newman, R.D.: Searches for anomalous long-range forces. In: Fackler, O., Tran Thanh Van, J. (eds.), pp. 599–606. Editions Frontières, Gif sur Yvette (1987)

Newman, R., Graham, D., Nelson, P.: A 'Fifth Force' search for differential acceleration of lead and copper toward lead. In: Fackler, O., Tran Thanh Van, J. (eds.), pp. 459–472. Editions Frontières, Gif sur Yvette (1989)

Niebauer, T.M., McHugh, M.P., Faller, J.E.: Galilean test for the Fifth Force. Phys. Rev. Lett. **59**, 609–612 (1987)

Nieto, M.M., Goldman, T., Hughes, R.J.: Phenomenological aspects of new gravitational forces. I. Rapidly rotating compact objects. Phys. Rev. D **36**, 3684–3687 (1987a)

Nieto, M.M., Goldman, T., Hughes, R.J.: Phenomenological aspects of new gravitational forces. II. Static planetary potentials. Phys. Rev. D **36**, 3688–3693 (1987b)

Nieto, M.M., Macrae, K.I., Goldman, T., Hughes, R.J.: Phenomenological aspects of new gravitational forces. III. Slowly rotating astronomical bodies. Phys. Rev. D **36**, 3694–3699 (1987c)

Nobili, A.M., Milani, A., Farinella, P.: Testing Newtonian gravity in space. Phys. Lett. A **120**, 437–441 (1987)

Nussinov, S.: Further tests and possible interpretations of a suggested new vectorial interaction. Phys. Rev. Lett. **56**, 2350–2351 (1986)

Paik, H.J.: Terrestrial experiments to test theories of gravitation. In: MacCallum, M.A.H. (ed.) General Relativity and Gravitation, pp. 388–396. Cambridge University Press, New York (1986)

Parker, R.L.: A non-non-Newtonian explanation for the results of certain recent gravity experiment. EOS **69**, 1046 (1988)

Parker, R.L., Zumberge, M.A.: An analysis of geophysical experiments to test Newton's law of gravity. Nature **342**, 29–32 (1989)

Peccei, R.D., Sola, J., Wetterich, C.: Adjusting the cosmological constant dynamically: cosmons and a new force weaker than gravity. Phys. Lett. B **195**, 183–190 (1987)

Pimental, L.O., Obregon, O.: A scalar-tensor theory and the new interaction. Astrophys. Space Sci. **126**, 231–234 (1986)

Pusch, G.D.: A new test of the weak equivalence principle. Gen. Relat. Gravitat. **19**, 225–231 (1987)

Raab, F.J.: Search for an intermediate-range interaction: results of the Eöt-Wash. I experiment. In: Fackler, O., Tran Thanh Van, J. (eds.), pp. 567–577. Editions Frontières, Gif sur Yvette (1987)

Renner, J.: Kísérleti vizsgálatok a tömegvonzás és a tehetetlenség arányosságáról. Matematikai és Természettudományi Értesitö **53**, 542–568 (1935)

Rizzo, T.G.: Hyperphoton production in W-Boson decay. Phys. Rev. D **34**, 3519–3520 (1986)

Roll, P.G., Krotkov, R., Dicke, R.H.: The equivalence of inertial and passive gravitational mass. Ann. Phys. (N.Y.) **26**, 442–517 (1964)

Schastok, J., et al.: Newton's law of gravity modified? Celestial mechanical consequences. Phys. Lett. Λ **118**, 8 10 (1986)

Schwarzschild, B.: Reanalysis of old Eötvös data suggests 5th force . . . to some. Phys. Today **XX**, 17–20 (1986)

Silverman, M.P.: Satellite test of intermediate-range deviation from Newton's law of gravity. Gen. Relat. Gravitat. **19**, 511–514 (1987a)

Silverman, M.P.: On the search for an intermediate-range modification of the gravitational force. Europhys. Lett. **3**, 1–4 (1987b)

Speake, C.C., Quinn, T.J.: Beam balance test of weak equivalence principle. Nature **321**, 567–568 (1986)

Speake, C.C., Quinn, T.J.: Search for a short-range, isospin-coupling component of the Fifth Force with use of a beam balance. Phys. Rev. Lett. **61**, 1340–1343 (1988)

Speake, C.C. et al.: Test of Newton's inverse square law of gravity using the 300 m tower at Erie, Colorado: Newton vindicated on the plains of Colorado. In: Fackler, O., Tran Thanh Van, J. (eds.) New and Exotic Phenomena '90: Tenth Moriond Workshop. Editions Frontières, Gif sur Yvette (1990a)

Speake, C.C., et al.: Test of the inverse-square law of gravitation using the 300-m tower at Erie, Colorado. Phys. Rev. Lett. **65**, 1967–1971 (1990b)

Spero, R., et al.: Tests of the gravitational inverse-square law at laboratory distances. Phys. Rev. Lett. **44**, 1645–1648 (1980)

Stacey, F.D., Tuck, G.J., Moore, G.I.: Geophysical tests of the inverse square law of gravity. In: Fackler, O., Tran Thanh Van, J. (eds.), pp. 557–565. Editions Frontières, Gif sur Yvette (1987a)

Stacey, F.D., et al.: Geophysics and the law of gravity. Rev. Mod. Phys. **59**, 157–174 (1987b)

Stubbs, C.W.: Seeking new macroscopic interactions: an assessment and overview. In: Fackler, O., Tran Thanh Van, J. (eds.) New and Exotic Phenomena '90: Tenth Moriond Workshop, pp. 175–185. Editions Frontières, Gif sur Yvette (1990a)

Stubbs, C.W.: Testing the equivalence principle in the field of the earth: an update on the Eöt-Wash experiment. In: Fackler, O., Tran Thanh Van, J. (eds.) New and Exotic Phenomena '90: Tenth Moriond Workshop, pp. 225–232. Editions Frontières, Gif sur Yvette (1990b)

Stubbs, C.W., et al.: Search for an intermediate-range interaction. Phys. Rev. Lett. **58**, 1070–1073 (1987)

Stubbs, C.W., et al.: Eöt-Wash constraints on multiple Yukawa interactions and on a coupling to 'isospin'. In: Fackler, O., Tran Thanh Van, J. (eds.), pp. 473–484. Editions Frontières, Gif sur Yvette (1989a)

Stubbs, C.W., et al.: Limits on composition-dependent interactions using a laboratory source: is there a 'Fifth Force' coupled to isospin? Phys. Rev. Lett. **62**, 609–612 (1989b)

Suzuki, M.: Bound on the mass and coupling of the hyperphoton by particle physics. Phys. Rev. Lett. **56**, 1339–1341 (1986)

Talmadge, C., Fischbach, E.: Phenomenological description of the Fifth Force. In: Fackler, O., Tran Thanh Van, J. (eds.), pp. 413–427. Editions Frontières, Gif sur Yvette (1988)

Talmadge, C., Aronson, S.H., Fischbach, E.: Effects of local mass anomalies in Eötvös-type experiments. In: Tran Thanh Van, J. (ed.), pp. 229–240. Editions Frontières, Gif sur Yvette (1986)

Thieberger, P.: Hypercharge fields and Eötvös-type experiments. Phys. Rev. Lett. **56**, 2347–2349 (1986)

Thieberger, P.: Search for a substance-dependent force with a new differential accelerometer. Phys. Rev. Lett. **58**, 1066–1069 (1987a)

Thieberger, P.: Search for a new force. In: Fackler, O., Tran Thanh Van, J. (eds.), pp. 579–589. Editions Frontières, Gif sur Yvette (1987b)

Thieberger, P.: Thieberger replies. Phys. Rev. Lett. **62**, 810 (1989)

Thodberg, H.H.: Comment on the sign in the reanalysis of the Eötvös experiment. Phys. Rev. Lett. **56**, 2423 (1986)

Thomas, J., Vogel, P.: Testing the inverse-square law of gravity in Boreholes at the Nevada test site. Phys. Rev. Lett. **65**, 1173–1176 (1990)

Thomas, J., Vogel, P., Kasameyer, P.: Gravity anomalies at the Nevada test site. In: Fackler, O., Tran Thanh Van, J. (eds.), pp. 585–592. Editions Frontières, Gif sur Yvette (1988)

Thomas, J., et al.: Measured free air gradients do not agree with model gravity gradients at the Nevada test site. In: Blair, D.G., Buckingham, M.J. (eds.) Proceedings of the Fifth Marcel Grossman on General Relativity, pp. 1573–1576. World Scientific, Singapore (1989)

Tran Thanh Van, J. (ed.): Progress in Electroweak Interactions, Proceedings of the Lepton Section of the Twenty-first Rencontre de Moriond, Workshop. Editions Frontières, Gif sur Yvette (1986)

Tuck, G.J.: Gravity gradients at Mount Isa and Hilton Mines. In: Abstracts of Contributed Papers, Twelfth International Conference on General Relativity and Gravitation, Boulder (1989)

Vecsernyes, P.: Constraints on a vector coupling to Baryon number from the Eötvös experiment. Phys. Rev. D **35**, 4018–4019 (1987)

Wetterich, C.: A new intermediate range scalar force? In: Fackler, O., Tran Thanh Van, J. (eds.), pp. 383–393. Editions Frontières, Gif sur Yvette (1988)

Whetton, J.T., et al.: Geophys. Prospect. **5**, 20 (1957)

Zumberge, M., Parker, R.: Newton gravitational constant. Science **238**, 1026–1027 (1987)

Zumberge, M., et al.: Results from the 1987 Greenland G experiment. EOS **69**, 1946 (1988)

Zumberge, M., et al.: Submarine measurement of the Newtonian gravitation constant. Phys. Rev. Lett. **67**, 3051–3054 (1991)

Chapter 3
Discussion

In this history I have examined the Fifth Force hypothesis from its origins, through its proposal and further investigation by other scientists, to its ultimate rejection by the physics community. These are what philosophers of science have called the contexts of discovery, of pursuit, and of justification. In previous work (Franklin 1990) I have argued that science follows an "evidence" model in which questions of theory choice, confirmation, and refutation are decided on the basis of valid experimental evidence.[1] I have applied this model to various episodes in the history of science, including the discoveries of parity violation (the violation of left-right symmetry in nature) and of CP violation (combined parity and particle–antiparticle symmetry violation) and argued that this evidence model applies to the context of justification. I believe that this history has not only provided us with another illustration that the evidence model works in the context of justification, but it has also allowed us to examine the contexts of discovery and pursuit and to investigate the role that evidence may play in these contexts.[2]

I will begin this discussion, however, with the ultimate fate of the Fifth Force, and the context of justification. It seems clear from the history presented that the conclusion that the Fifth Force[3] does not exist was based on an overwhelming preponderance of experimental evidence. It is also quite clear that the process was

---

[1] I have also argued that there is an epistemology of experiment, a set of strategies for arguing for the validity of experimental results.

[2] It may not always be possible to clearly separate these contexts, but I believe that we should do so where it is possible. I believe it adds clarity to the discussion. For example, in the discussion of Fischbach's work on the development of the Fifth Force hypothesis, I discussed it as the context of discovery because there was no hypothesis being investigated. It might also have been discussed as the pursuit of a solution to the problem of CP violation and its possible connection to gravity.

[3] I refer here to the original proposal of a force with a strength approximately 1 % that of gravity and a range of about 100 m.

© Springer International Publishing Switzerland 2016
A. Franklin, E. Fischbach, *The Rise and Fall of the Fifth Force*,
DOI 10.1007/978-3-319-28412-5_3

far more complex than "Man proposes, Nature disposes."[4] One has to deal with the fallibility of experiment, of theory, and of the comparison between experiment and theory.

We have seen that it was not clear what was being proposed, or, perhaps more accurately, that as things developed there were several proposals. The original suggestion of the Fifth Force did not include the effect of local mass asymmetries. This implication was quickly realized by both the original authors and by others, and the calculations were corrected.[5] As conflicting experimental evidence regarding the force appeared, more complexity was added to the theoretical model, including its possible dependence on quantities other than baryon number, such as isospin, and more complex distance dependence in the form of multicomponent Yukawa models or pure exponential potentials.

We have also seen that it was not immediately apparent how Nature was disposing. The two initial experimental results, those of Thieberger and of Eöt-Wash, gave conflicting results, one favoring the existence of the Fifth Force and one opposed. These were followed shortly thereafter by Boynton's "marginally observed" (his words) positive result. The subsequent history seems to be an illustration of one way in which the scientific community deals with conflicting experimental evidence.[6] Rather than making an immediate decision as to which were the valid results, this seemed extremely difficult to do on methodological or epistemological grounds, the community chose to await further measurements and analysis before coming to any conclusion about the evidence. The torsion-balance experiments of Eöt-Wash and Boynton were repeated by others including Fitch, Cowsik, Bennett, and Newman and by Eöt-Wash and Boynton themselves. These repetitions, in different locations and using different substances, gave consistently negative results. In addition, Bizzeti, using a float apparatus similar to that of Thieberger, also obtained results showing no evidence of a Fifth Force. There was an overwhelming preponderance of evidence against the existence of a Fifth Force, particularly against any possible composition dependence of such a force.

There is an interesting methodological point here. Colin Howson and I (Franklin and Howson 1984) have previously argued that when experimental results agree, then "different" experiments provide more support for a hypothesis than repetitions of the same experiment.[7] In this case the hypothesis would be: "The Fifth Force exists." When results disagree, however, then it may be the differences in possible backgrounds (effects that might mimic or disguise the real effect) and systematic

---

[4]For other illustrations of this see the discussions of the interaction of experiment and theory in the case of weak interactions and atomic parity violation in Franklin (1990).

[5]This refinement of the theoretical model allowed more sensitive experimental tests of the Fifth Force to be both designed and implemented.

[6]Another method will be discussed later.

[7]"Different" experiments are classified by the theory of the apparatus. See Franklin and Howson (1984) for details.

errors that account for the discordant results.[8] Thus, in principle, the torsion balance experiments would probably not have been individually as effective in casting doubt on Thieberger's result as was Bizzeti's negative result with a similar apparatus. In reality, scientists made their own judgments on the quality and reliability of the different experiments. Had Bizzeti agreed with Thieberger then one might well have wondered whether or not there was some systematic difference between torsion-balance experiments and float experiments that gave rise to conflicting results.[9] Fortunately, that did not occur. There is, in fact, no explanation of either Thieberger's or of Boynton's (Index I), presumably incorrect, results. The scientific community has chosen, I believe quite reasonably, to regard the preponderance of negative results as conclusive.[10] It was this preponderance of evidence along with the negative distance-dependence geophysical measurements that led to the conclusion that the Fifth Force did not exist.

Fallibility can also extend to both theoretical calculation and to theory–experiment comparison. This is what happened in the case of the geophysical measurements of Stacey, of Eckhardt, and of the Greenland group. No one has suggested that the measured values of gravity were incorrect, but rather than the theoretical values used for comparison were wrong. One found, somewhat surprisingly given the long history of such gravitational calculations, that more care was needed in these calculations. As Bartlett, and others, pointed out, one needed careful considerations of the local terrain as well as detailed surface gravity measurements in order to make accurate upward- or downward-continuation calculations. Failure to do this accounted for the gravitational anomalies originally reported by Stacey and Eckhardt.[11] Similarly, Parker and Zumberge showed that the gravity anomalies found by the Greenland group might be explained by local density variations and suggested that such variations might also account for the Stacey and Eckhardt anomalies. These instances of fallibility also show us the corrigibility

---

[8]It could also be the case that both experimental results were correct, or they could both have been wrong.

[9]In the 1930s it was found that experiments on beta decay using thick and thin sources, respectively, gave consistently different results, and a systematic error was later found in the thick target experiments. (See Franklin (1990, Chap. 1) for details.) There is a similar contemporary problem that has not yet been resolved. Scientists using one type of detector have found suggestive evidence for a neutrino with a mass of $17\,\text{keV/c}^2$. No such evidence appears when another type of detector is used. The question remains whether one of these results is an artifact of the detector used and the other a valid result, and if so, which one is valid.

[10]It is a fact of experimental life that experiments rarely work when they are initially turned on and that experimental results can be wrong, even if there is no apparent error. It is not necessary to know the exact source of an error in order to discount or to distrust a particular experimental result. It's disagreement with numerous other results can, I believe, be sufficient.

[11]In the intermediate range (of the order of hundreds of meters) the local gravity measurements near Eckhardt's North Carolina tower were taken at too high an average elevation because of the local swampy ground. Near Stacey's Australian mine site, the local measurements in this range were too low because they thought that the mineshafts would be located in valleys. This also explained the difference in sign of the two observations.

of science. The errors in the theoretical calculations and in the experiment–theory comparison were corrected. In the case of the discordant experimental results more experiments were performed to decide the issue.

One might worry that only those results that disagreed with accepted gravitational theory were subjected to this careful scrutiny and this is, to a certain extent, true. The scrutiny and criticism given by others to the anomalous geophysical measurements, to the results of Thieberger and Boynton, and to the original reanalysis of the Eötvös experiment, seems to have exceeded that given, at least publicly, to the experiments that agreed with accepted gravitational theory and argued against the Fifth Force. This overlooks, however, the internal scrutiny and criticism given to the experimental results by the experimenters themselves,[12] who have an interest in presenting correct results, and also overlooks the considerable informal criticism even between groups whose results agreed both with each other and with accepted theory. I was present during several such discussions during the 1989 and 1990 Moriond workshops and can testify to the rigor and detail of such criticism.[13] In addition, the anomalous results were in conflict with all of the previous experimental evidence that supported accepted gravitational theory, and were, therefore, more likely to be incorrect. There was, of course, no guarantee that this would be the case, and thus, the criticism of all the results and the repetition of the experiments.

Along these lines Pickering (1981a, 1984a,b) has raised several interesting questions concerning experimental results and their use as evidence. Pickering suggests that experimental results may be accepted either because of the future utility of such results for the practice of science, or because they fit in with existing community commitments. Although these are often related, they are not identical. More generally, Pickering is a representative of the constructivist position, in which scientific decisions are based on interests, which may be social, either class or religious, or more narrowly professional, such as future utility or recycling of theoretical or experimental expertise.[14] The constructivist view is that evidence cannot,[15] and does not, decide these issues so there must be other reasons for the decision and these are the interests of those involved. This view also denies that

---

[12]For an in-depth look at an illustration of this see Galison (1987, Chap. 4).

[13]There is an advantage for the historian or philosopher of science to be present while the science is being done and discussed. There is also an obvious danger that one will identify too closely with the participants. In this case I believe the danger has been avoided because I had no preference as to whether or not the Fifth Force actually existed. The history would be just as interesting and instructive in either case.

[14]For a useful introduction to this constructivist position see Pinch (1986, Chap. 1) and also comments in Franklin (1990, Chap. 8).

[15]The constructivists depend here on philosophical arguments such as the underdetermination of theory by evidence, the incommensurability and the theory ladeness of observation, and the Duhem-Quine problem (the problem of assigning blame when a theory is apparently refuted). For a discussion of these issues see Nelson (1994) and Franklin (1990, Chap. 7).

there is an epistemology of experiment, a set of strategies that can provide good reasons for belief in the validity of experimental results.

In his recent book, *Constructing Quarks* (1984a), Pickering discussed the early experiments on atomic parity violation, which were anomalous for the Weinberg–Salam (W–S) unified theory of electroweak interactions. These experiments, performed at Oxford University and the University of Washington and published in 1976 and 1977, measured the parity non-conserving optical rotation in atomic bismuth. The results disagreed with the predictions of the Weinberg–Salam theory, which had other experimental support. Another experiment, performed in 1978 at the Stanford Linear Accelerator on the scattering of polarized electrons from deuterons, confirmed the theory. Pickering regards the Oxford and Washington experiments as mutants slain by the SLAC experiment. By 1979 the Weinberg–Salam theory was regarded as established by the high-energy physics community despite the fact that as Pickering recounts (1984a, p. 301), "there had been no *intrinsic* change [emphasis in original] in the status of the Washington–Oxford experiments." In Pickering's view (1984a, p. 301), "particle physicists *chose* [emphasis in original] to accept the results of the SLAC experiment, *chose* to interpret them in terms of the standard model; (rather than some alternative which might reconcile them with the atomic physics results), and therefore *chose* to regard the Washington–Oxford experiments as somehow defective in performance or interpretation." The implication seems to be that these choices were made so that the experimental evidence would be consistent with the accepted standard model, and that there were not good, independent reasons for that choice.

Pickering's explanation of this is an "interest" model in which agreement with accepted theory provided both experimentalists and theorists with more work to do than would, presumably, have been available had the results disagreed with accepted theory. In discussing another episode, the discovery of weak neutral currents, Pickering expresses a similar view (1984b, p. 87): "Quite simply, particle physicists accepted the existence of the neutral current because they could see how to *ply their trade more profitably* [emphasis added] in a world in which the neutral current was real."

I have argued elsewhere (Franklin 1990, Chap. 8) that Pickering's historical analysis of the episode of atomic parity violation experiments is wrong. In my view, the reason for the decision of the scientific community to accept the W–S theory of the basis of the SLAC experiment was because that experiment had far greater evidential weight than did the early Washington-Oxford results. It was an extremely carefully done and checked experiment, while the early atomic parity violation experiments were quite uncertain and had admitted, large systematic uncertainties, of the same order of magnitude as the predicted effects. The physics community chose to accept the carefully done result and to await further work on the atomic parity experiments. I note here that subsequent work on these experiments did, in fact, confirm the W–S theory. This is another way in which the physics community deals with conflicting evidence, by actually evaluating how well the experiments were done and what their evidential weight was. (For differing views of that episode see Ackermann (1991), Lynch (1991), and Pickering (1991).)

Pickering (1981b) has presented another case to support his view that prospects for future practice are important considerations in the acceptance of a theory, by which he means its justification, rather than its pursuit. At the time of the discovery of the ψ particle there were two competing theoretical explanations of the phenomenon. One, the charm model, made explicit predictions of experimental results, while the other, the color model made no such precise predictions. There was also experimental evidence which gave partial, although not complete, support to the charm view. In addition, the charm model fitted in better with an existing theoretical program on gauge theories. Pickering regards this desire to recycle one's theoretical expertise as the sole explanation of this episode and discusses it as the basis of the justification of the charm model. Given the specificity of the predictions, the partial confirmation, and the existing skills of physicists, it is no great surprise that most physicists in the field chose to work on the charm model.[16] I agree with Pickering that the opportunity for future work is an important motivation for scientists. I would even agree that other interests such as career advancement, recycling of expertise, etc., may also be important motives. I am suggesting, however, that these considerations are important in the context of pursuit and not in the context of justification. It is this failure to distinguish between the contexts of pursuit and justification that, I believe, leads Pickering astray. I shall return to the discussion of the context of pursuit later.

Let us see, however, whether or not a Pickeringesque analysis can be applied to the case of the Fifth Force experiments. In the case of the conflict between Thieberger's result and that of the Eöt-Wash group the physics community, at least those actively working in the field, chose to await further experimental work before deciding which of these two results was correct. (Some theorists, as we have seen, worked on scenarios that allowed both results to be correct.) Here, unlike the case of W–S theory and the atomic parity and SLAC experimental results, there was no clear argument for the methodological superiority or greater evidential weight of one experiment over the other, and hence the lack of a decision. This does seem to support my view that scientists do evaluate the reliability and evidential weight of experimental results.[17] Precisely because epistemological analysis was unavailing, further work was needed. Thieberger (1989), himself, in the light of the subsequent failure find positive results in agreement with his, believes that there must be an undiscovered systematic error in his experiment. Similarly, the anomalous geophysical results were not dismissed in favor of those supporting accepted theory until after further critical analysis had uncovered problems in the experiment–theory comparisons, and other experiments, which corrected these problems, had given results that agreed with accepted theory. Had these new experiments also disagreed

---

[16]Ultimately experiment provided evidence of the superiority of the charm model.

[17]One might, of course, claim that no such evaluation took place. After all, there was very little published criticism, but this overlooks the large amount of unpublished criticism such as that which took place at the Moriond Workshops.

with accepted theory the anomaly would have been confirmed, and would obviously have become more serious.

It might also be argued that scientists accepted these explanations in order to get results in agreement with existing theory. As Pickering states (1981a, p. 236), "scientific communities tend to reject data that conflict with group commitments and, obversely, to adjust their experimental techniques and methods to 'tune in' on phenomena consistent with those commitments." There is evidence from the history of measurements of the gyromagnetic ratio of the electron (Galison 1987, Chap. 2) and Millikan's oil drop experiment (Franklin 1986, Chap. 5) that the theoretical presuppositions of some experimenters influenced the results they presented. The subsequent history showed, however, that for important quantities there are usually numerous repetitions of the measurements and that the results later agreed upon did not necessarily agree with the presuppositions. In the case of the gyromagnetic ratio of the electron the value later agreed upon for the g factor was 2, in disagreement with the theoretical prediction that it would be one. In Millikan's case, his expectation that charge was quantized was confirmed, as was his numerical value for the charge on the electron.[18]

There seems to be no evidence of this "tuning in" phenomenon in the Fifth Force episode. Both Eckhardt and Stacey admitted publicly that analysis errors had been made, certainly not something that enhances one's reputation. As Peter Galison said (1987, p. 2): "Research reputations will hinge on their [the experimenters'] judgment that they have adduced adequate evidence." I note also that it took some time, and considerable criticism, before Stacey and Eckhardt were convinced of the correctness of the criticisms. They believed, at least for a time, that their comparisons between experiment and theory were correct, and that there was a discrepancy between experiment and theory.

It is also not clear why the interest of scientists in subsequent practice should require results that agree with existing, accepted theory. I believe that Pickering's belief that it does, stems, in part, from a theory-dominated view of experiment. In his view the only important role of experiment seems to be that of testing theory. He implies that without theory as a guide, experimenters would have little, if anything, to do. This is an impoverished view of experiment. Not only does experiment often have a life of its own (a point I shall return to later), but it can also call for a new theory, give hints and suggestions as to the form of that theory, or acquire data that any theory of the phenomena will have to explain.

The history of physics has shown us that even very strongly held beliefs such as parity conservation and CP symmetry can be overthrown by experimental evidence (See Franklin (1986, Chaps. 1 and 3) for details of these episodes.)[19] These results certainly called for new theories and led to considerable experimental

---

[18]Millikan's value for e actually differs from the modern value. The difference can be explained by differences in the values used for the viscosity of air.

[19]The experimental refutation of such strongly held beliefs can, in fact, have beneficial career effects. Thus, Cronin and Fitch, the experimenters who found CP violation, won the Nobel Prize.

and theoretical work. The experiments on the Meissner effect, the exclusion of magnetic fields from the interior of a superconductor, gave hints toward a theory of superconductivity and experiments on beta decay during the 1930s, 1940s, and 1950s provided evidence concerning the mathematical form of the theory of weak interactions. (See Franklin (1990, Chaps. 1–5) for details.) In addition, experiments on strongly interacting particles, including the discovery of new particles and resonances, during the 1960s and 1970s, were done without any accepted theory of the phenomena, as were the experiments on atomic spectra during the nineteenth and early twentieth centuries. These experiments provided data for a new theory of the phenomena to explain.

In the case of weak neutral currents and atomic parity violation there was no plausible competing theory (a hybrid model had been ruled out experimentally).[20] The Fifth Force hypothesis as an alternative, or perhaps, as an addition to gravitational theory, certainly seemed to provide useful employment for both theorists and experimentalists.[21] As we have seen, it even attracted new physicists to the field. Experimenters investigated both the composition dependence and the distance dependence of the presumed force, while theorists offered models to both explain the force and the newly acquired experimental data. The overthrow of an existing theory is an opportunity for theorists to invent a new theory and an opportunity for experimenters to find data that will assist in the search for such a replacement or possibly confirm or refute it. This is what happened in the case of the Fifth Force.

In summary, I believe that the evidence model fits the context of justification in the Fifth Force episode. I don't believe that one can make a case for a Pickeringesque analysis.

Someone might also claim that experimentalists will tend to get results that agree with those of previous experiments whether or not those results agree with existing theory. This is an experimental, rather than a theoretical, bandwagon effect. I believe that this bandwagon effect is a possible danger in science. I have, however, argued previously (Franklin 1986, Chap. 8) that it is not a real problem in the practice of science. Although experimenters may very well end their search for systematic errors when they get a result in agreement with previous results, one that is therefore more likely to be accepted, they have an even greater interest in presenting correct results. The history of physics has shown that even very well confirmed measurements may change by far more than expected given the cited experimental uncertainties. This is clearly indicated in the history of the measurements of $|\eta_{+-}|$, the CP violating parameter in K^0 decay, shown in Fig. 3.1. The measured value of $|\eta_{+-}|$ changed dramatically in 1973 and it seems clear that the first of the later group

---

Similarly, Lee and Yang, theorists who suggested parity violation and its experimental tests, also won the Nobel Prize.

[20]This hybrid model allowed both the neutrino results at high energies and the low energy atomic parity violation results to be correct.

[21]Because the Fifth Force involves the exchange of a different particle than that exchanged in existing gravitational theory one might very well consider it a new force.

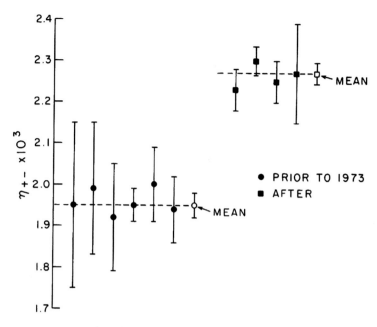

Fig. 3.1 Measurements of $|\eta_{+-}|$ in the order of their publication

of experimenters had sufficient confidence in their own work to report a result that disagreed with the previous world average by eight standard deviations,[22] a very unlikely result if they are measurements of the same quantity. (The probability of such a difference is 1.24×10^{-15}.) They were also willing to disagree with the results of six other experiments.

One might also ask which result the later post-1973 experimenters should have selected to agree with, the earlier consistent measurements or the latest discordant one. Without clairvoyance they had no guidance. They could, of course, evaluate all of the previous measurements and then choose to agree with those they thought most reliable. This assumes that they have not used the same kind of evaluation on their own experiment, and that they do not have sufficient confidence in their own work to report their result, unlike Steinberger's group, the earliest discordant post-1973 experiment. There was then, and is even now, no theoretical prediction of the value of $|\eta_{+-}|$, so theory provided no guidance. What is the poor experimenter to do?

Similarly, in the measurements on the Fifth Force, subsequent experimenters had no methodological or epistemological guide as to whether the Eöt-Wash or the Thieberger result was correct. They both appeared to be carefully done. Absent such a guide they were forced to rely on their own best judgment on the quality of their

---

[22]This is, in fact, the second point plotted in the later group of results. Although it was reported earlier at a conference it was the second published result.

own experiment, something they were likely to do in any event.[23] Agreement with accepted theory didn't provide any assistance either because, as discussed earlier, an experiment that disagrees with or overthrows existing accepted theory can have beneficial effects both for science, and for scientists. Although experimental results may be fallible, scientists do offer arguments for the validity of their measurements and these arguments are critically examined by the scientific community.

I now turn to the role that evidence may play in the contexts of discovery and pursuit. I discussed earlier that it is important to try to distinguish carefully between the contexts of pursuit and justification. The failure to do so is illustrated in a recent comment on the Fifth Force by Anderson (1992). Anderson has attempted to align Bayesian confirmation theory against the original Fifth Force hypothesis (p. 9): "From the Bayesian point of view, it is not clear that even the very first reexamination of Roland von Eötvös' results actually supported the fifth force, and it's very likely that none of the 'positive' results were outside the appropriate error limits." I believe that Anderson's attempt misses the point. He fails to distinguish between the evidence necessary for believing in a hypothesis and that needed for investigating it further. At the time of the original publication the question was not one of belief, but rather whether or not one should pursue the possibility of such a force.

Anderson presents us with few details of either his Bayesian analysis or of Bayesianism itself. Briefly summarized, Bayesian confirmation theory includes the view that scientists have degrees of belief and that these can be represented by probabilities. (For a detailed account of Bayesianism see Howson and Urbach (1989).) The fundamental mathematical theorem of Bayesianism is Bayes's Theorem, which is

$$P(h/e) = P(e/h)P(h)/P(e) ,   \qquad (3.1)$$

where $P(h/e)$ is the probability of h given e, similarly for $P(e/h)$, and where $P(h)$ and $P(e)$ are the prior probabilities of h and e, respectively. All probabilities are relative to some store of background knowledge. Thus, Bayes's Theorem tells us how we should change our beliefs in the light of evidence. It does not, however, tell us what our initial beliefs, or prior probabilities, should be. Because $P(e)$ is always less than one, and $P(e/h) = 1$ when h entails e, this gives us the intuitively appealing result that observation of evidence entailed by a hypothesis should strengthen our belief in that hypothesis. This is, I believe, one of the major reasons for favoring Bayesianism.

One of the problems with this view is the assigning of the prior probabilities. As far as I know, there is no satisfactory way of assigning them objectively. The best we can do is to use subjective probabilities, which allows different scientists to

---

[23]I am not denying that experimenters consider previous results when they are analyzing their own experiments, but rather that this consideration alone does not, in the long run, lead to convergence. See the histories of the measurements of other quantities given in Franklin (1986, Chap. 8).

make different judgments about the plausibility of a hypothesis.[24] I believe that this inclusion of the possibility of differing probability assignments is another strength of the Bayesian view. It is certainly descriptively accurate.[25] As we have seen, scientists do not make the same judgments about the plausibility of a hypothesis. Nevertheless, I believe that in the actual practice of science the initial assignment of priors will not differ by so much that scientists will not converge, given a reasonable amount of evidence, to the best supported hypothesis, even though they may not agree on the value of the posterior probability, $P(h/e)$.[26] (See Franklin (1990, Chap. 6) for a short introduction to Bayesianism and its application to both confirmation theory and to an epistemology of experiment.)

Although most of us who regard Bayesianism as a fruitful way of looking at both science and its philosophy rarely, if ever, actually make use of numerical values for the probabilities, let us see how one might reasonably apply Bayesianism to the episode of the discovery of the Fifth Force, using such numerical values. If all goes well, our conclusions will not be sensitive to the values chosen.

Let us suppose that the only two competing hypotheses were h_1, Newton's Law of Gravitation, and h_2, Newton's Law plus the Fifth Force.[27] Let e be the results of the reanalysis of the Eötvös experiment. Recall that Fischbach et al. obtained a fit $\Delta k = (5.65 \pm 0.71) \times 10^{-6} \Delta(B/\mu) + (4.83 \pm 6.44) \times 10^{-10}$, where Δk is the fractional change in gravitational acceleration, B is the baryon number of the substance, and μ is the mass of the substance in units of the mass of hydrogen. The linear coefficient is eight standard deviations from zero.

If h_1 and h_2 are the only two competing hypotheses then

$$P(e) = P(e/h_1)P(h_1) + P(e/h_2)P(h_2) , \qquad (3.2)$$

and

$$P(h_1) + P(h_2) = 1 . \qquad (3.3)$$

When the Fifth Force was proposed in early 1986 it seems fair to say that Newtonian gravity (or General Relativity, its modern successor) was very strongly believed and that, although Fujii and others were working on modifications of the Brans–Dicke

---

[24]These probabilities are sometimes explicated by betting behavior, what the scientist thinks would be fair odds on the truth of the hypothesis. This should not be interpreted as real betting behavior because, as illustrated below, someone may not be willing to bet at what they actually consider to be fair odds. See Howson and Urbach (1989, Chaps. 1–2) for details.

[25]Such differing estimates of probability also lead to more hypotheses being investigated.

[26]Earman (1992) shows that for distinguishable hypotheses, those for which there are different empirical consequences, there are convergence of opinion results for Bayesianism.

[27]I am really assuming here that they were the only two hypotheses with any significant prior probability. This seems reasonable. By this time experiment had eliminated the Brans–Dicke theory as a serious competitor.

theory, the Fifth Force hypothesis was neither widely nor strongly believed. Let us set $P(h_1) = 0.999$ and $P(h_2) = 0.001$, although as we shall see our conclusions do not depend critically on these values. The assignments do seem reasonable, given the theoretical work by Fujii and others, along with the suggestive experimental evidence provided by the geophysical gravitational anomalies and the energy dependence of the $K^0 - \overline{K}^0$ parameters. If one takes the eight standard deviation difference from zero as correct then $P(e/h_1) = 1.24 \times 10^{-15}$, because h_1 predicts a value of zero and that is the probability of an eight standard deviation effect. $P(e/h_2) = 1$.

We can write

$$P(h_1/e) = P(e/h_1)P(h_1)/P(e)$$

$$= P(e/h_1)P(h_1)/[P(e/h_1)P(h_1) + P(e/h_2)P(h_2)]$$

$$= (1.24 \times 10^{-15})(0.999)/(1.24 \times 10^{-15})(0.999) + 1(0.001)$$

$$= (1.24 \times 10^{-15})/(0.001) = 1.24 \times 10^{-12} \, . \tag{3.4}$$

This implies that one's belief in Newton's law should be close to zero in the light of the results of the reanalysis of the Eötvös experiment. Of course, no one would, or should, take this analysis seriously. Historically, the strong belief in Newton's law remained virtually unchanged. The reason for this was that no one seriously believed that the eight standard deviation effect reported by Fischbach was correct, although as we have seen several people thought it was intriguing. Given the selection of data, the criticisms offered by others, and the uncertainty in the analysis and estimation of systematic errors for an experiment performed more than 70 years earlier, this skepticism seems reasonable.

Perhaps a more reasonable estimate of the effect seen in the reanalyzed data was that it was a two or three standard deviation effect, roughly comparable to the energy dependence of the $K^0 - \overline{K}^0$ parameters and the geophysical gravitational anomalies reported earlier. In the case of a three standard deviation effect, $P(e/h_1) = 0.0027$ and $P(e/h_2) = 1$. Using (3.4), we find

$$P(h_1/e) = (0.0027)(0.999)/[(0.0027)(0.999) + 0.001] = 0.73 \, ,$$

which also gives $P(h_2/e) = 0.27$, in rough agreement with Newman's estimate of the plausibility of the Fifth Force after reading the original paper. If one changes the Eötvös effect to two standard deviations (recall that my own reanalysis of the Eötvös data, using his published values, gave a fit to a straight line which was within two standard deviations of zero—see Fig. 1.11), we find that $P(h_1/e) = 0.98$ and $P(h_2/e) = 0.02$ for this assignment of the prior probabilities.

Our conclusion that Newtonian gravity remained strongly believed whereas belief in the Fifth Force was weak, although not zero, does not depend sensitively on the values of the probabilities chosen for $P(e/h_1)$ in these last two cases. Our conclusions are also reasonably robust under changes in the prior probability of the Fifth Force, $P(h_2)$. For $P(h_2) = 0.0001$ we find that for a three standard deviation Eötvös effect the posterior probabilities $P(h_1/e)$ and $P(h_2/e)$ are reduced to $P(h_1/e) = 0.96$ and $P(h_2/e) = 0.04$. For a two standard deviation effect, or much weaker evidence, $P(h_1/e)$ becomes 0.998 with a corresponding reduction for $P(h_2/e)$.[28]

The point of this analysis is not in the numerical values obtained, but rather that for reasonable estimates of the probabilities of both the Fifth Force and of the evidence, the Fifth Force had some plausibility. Bayesianism does seem to give a reasonable description of what actually happened.[29] I am not suggesting that any member of the physics community actually performed such calculations, but rather that they made qualitative judgments along these lines.[30] Of course, other members of the physics community might very well have given the Fifth Force hypothesis a much lower probability than 0.001 and regarded the evidence as far more uncertain, as for example Glashow,[31] although, as shown above, our conclusions are not very sensitive to these probability assignments.

Anderson has not distinguished between the contexts of pursuit and of justification. Although I disagree with the details of Anderson's analysis and with his conclusion concerning the pursuit of the Fifth Force, I am sympathetic to his attempt to apply Bayesianism to the practice of science and I agree with his conclusion that there was no strong evidence in support of the Fifth Force when it was originally proposed. I do think, however, that it was reasonable that a few members of the physics community thought it plausible enough to be worth further investigation.[32]

Anderson seems to regard the pursuit of speculative proposals as a waste of physicists's time. Perhaps this is because the Fifth Force turned out not to exist. One should remember, however, that parity nonconservation was regarded as highly unlikely by most of the physics community when it was first suggested (see Franklin (1986, Chap. 1) for details). For example, Feynman bet Norman Ramsey, who was planning an experiment to test parity nonconservation, \$50–\$1 that parity would be

---

[28]For $P(h_2) = 0.00001$ we find $P(h_1/e) = 0.9998$ and 0.996 for two and three standard deviation effects, respectively. $P(h_2/e)$ is reduced accordingly.

[29]Some critics might remark that I could always have chosen the prior probabilities to give the results I wanted. This is true, but I believe that the assignments I have made are reasonable given the fact that some scientists were indeed working on such theories when the hypothesis was offered, and the other evidence available at the time. I have, of course, already suggested that pursuit does not necessarily involve belief, but, even so, I think the calculations are reasonable.

[30]Scientists do make such judgments although they tend to be qualitative, i.e., large, small; or comparative, i.e., larger, smaller, etc.

[31]There were others who probably assigned the hypotheses a larger prior probability, such as Fischbach. I am not asserting that this analysis applies to any single member of the physics community, but only that it represents reasonable estimates of community beliefs.

[32]I am here associating plausibility with probability, although I do not believe that there is a uniform threshold of probability for pursuit, but rather that different scientists will make different judgements as to whether or not a hypothesis is plausible enough to be worth pursuing. Recall the earlier discussion on interest, plausibility, and ease of test in the context of pursuit.

conserved. When the experimental results were announced, Feynman paid.[33] In a
letter to Feynman, Ramsey noted (private communication): "You may recall that I
originally offered you the bet at 1000 to 1 odds and you reasonably declined on
the grounds that you would not bet on anything at 1000 to 1 odds."[34] Wolfgang
Pauli, another noted physicist, also had strong doubts. He wrote (quoted in Bernstein
(1967, p. 59)): "I do not believe the Lord is a weak left-hander, and I am ready to
bet a very large sum that the experiments will give symmetric results."[35]

Another point, discussed earlier, and illustrated by this history, is that investigating or pursuing a hypothesis does not require a strong belief in its truth. The decision
to pursue an investigation seems to depend on a weighting of at least three factors;
the interest of the hypothesis, its plausibility, and its ease of test. The latter may very
well depend on the expertise of the experimenters and the availability of equipment
and/or funding.

Belief in the truth of a hypothesis or in experimental results is also not a
requirement for further theoretical work. Thus, in the episode of atomic parity
violation discussed briefly earlier, Dydak, in attempting to calculate the electron–
quark coupling constants, chose to accept both the W–S theory and the experimental
results that agreed with it, rather than those that disagreed, although he admitted
that this choice could not be justified at the time. He needed to make some
assumptions concerning both the evidence and the theoretical model in order to
make any calculations at all. Nor is a real commitment to the truth of a hypothesis a
requirement for proposing it. In the case of CP violation, Bell and Perring remarked
(1964, p. 348): "Before a more mundane explanation is found, it is amusing to
speculate that it [CP violation] might be a local effect due to the dys-symmetry
of the environment, namely the local preponderance of the matter over antimatter."
Lee was a coauthor of a proposal similar to that of Bell and Perring, coauthor of
another alternative explanation of CP violation, and the author of a model to *avoid*
CP violation. Clearly, he was just speculating, and not committed to the truth of all
three proposals.

This pragmatic view also applies to extending work on a theory. At the moment,
string theory has no empirical consequences that can be tested by experiment.[36]
Nevertheless, many physicists are pursuing it because of its future promise as a
"theory of everything," a phrase often used to describe the theory. The theory also
has nice mathematical properties. It is a finite theory and thus can have experimental
tests, even if none are calculable or possible now, and it also allows theorists to
continue to use their expertise in gauge theory in this work.

---

[33]Ramsey never performed the experiment because his collaborator became involved in another
experiment that seemed more important at the time.

[34]I note that the real bet differed from what they thought were the fair odds.

[35]For those who might believe that authority and power play a major role in such decisions, I note
that Feynman, Ramsey, and Pauli all won the Nobel Prize in physics.

[36]At the moment, this is due to the difficulty of the theoretical calculations.

I am not suggesting that evidence is not important in the context of pursuit, but rather that it plays a different role than in the context of justification, in which I believe it is decisive. In the context of pursuit the evidence must only be strong enough to make the hypothesis plausible enough to merit further investigation. The Fifth Force hypothesis of Fischbach et al. (1986) differs little from that suggested earlier by Fujii. The reason it generated so much more work was that it was made more plausible by the three pieces of evidence presented, particularly the reanalyzed Eötvös data.[37] The extensive coverage in the popular press as well as the numerous seminars on the subject also helped.

This episode has also provided more evidence that experiment often has a life of its own. We have seen that the experiments on the Fifth Force, or rather similar experiments, are continuing even though the original Fifth Force hypothesis is no longer believed in. At least, in part, this is because the previous experience of the experimenters with the same type of apparatus has made them familiar with sources of background which might mimic or mask the effect and with sources of possible systematic errors. They can thus construct new apparatus, or modify existing apparatus, of greater precision and accuracy. This results in better experiments, and is also cost effective. We have also seen that interest in resolving an experimental discrepancy formed an important part of the motivation for further experiments. I do not wish to deny the role of theoretical context, that would be obviously incorrect, but I am suggesting that experiment does have other roles than testing theory, including the resolution of experimental discrepancies.

I would, however, like to emphasize that one important role of theory is that of acting as an "enabling theory" for experiment. (See the discussion in Galison (1987, Chap. 2).) The Fifth Force hypothesis not only motivated experimental work, but also provided an estimate of the size of the effects expected and thus had an important influence on experimental design. It also allowed experimenters to compare the relative size of the theoretically expected effects with those expected from background and to judge the feasibility of the experiment. Without such estimates an experiment may well be inconclusive. Thus, theory enters into the context of pursuit in helping to decide the ease or feasibility of an experimental test of the theory. I note here that even an incorrect hypothesis, as presumably the Fifth Force is, may act as an enabling theory. The criticism offered of the initial proposal also helped design more sensitive experiments by emphasizing the importance of local mass asymmetries and led to the hillside experiments.

---

[37]In the case of parity violation, the fact that the hypothesis that parity was not conserved solved the rather vexing $\Theta - \tau$ puzzle increased its plausibility considerably. See Franklin (1986, Chap. 1) for details.

The fact that the subsequent history has argued against the existence of the Fifth Force should not cause us to overlook the fact that the suggestion of that force was the result of a sequence of reasonable and plausible steps. This started with the Colella, Overhauser, and Werner measurement, Fischbach's attempt to connect CP violation and gravity, and the subsequent observation of suggestive energy dependence of the $K^0 - \overline{K}^0$ parameters. At the same time the work on the modifications of Newtonian gravity and the tantalizing results on the measurement of gravity in mines were proceeding. When these two strands were joined together it led the collaborators to reanalyze the original Eötvös experiment, where, again, a suggestive effect appeared. The suggestion of the Fifth Force then followed. There may not be a logic of discovery, but, at least in this case, it is not a totally mysterious process.

There is one rather intriguing puzzle. In retrospect, it appears that two of the three pieces of evidence that increased the plausibility of the Fifth Force, the geophysical gravity anomalies and the energy dependence of the $K^0 - \overline{K}^0$ parameters are incorrect. The geophysical anomalies have been explained by the failure to take local terrain into account properly. Subsequent work on K^0 mesons at much higher energies has left little or no trace of the energy dependence of the parameters (see Fig. 1.4). It appears that the tantalizing two or three standard deviation effects seen in K meson physics were merely statistical fluctuations.

What about the effect seen in the reanalysis of the Eötvös experiment? The effect originally reported by Fischbach was an eight standard deviation effect, if one believes the experimental uncertainties. A statistical fluctuation this large is quite improbable. Although later work (Fig. 2.27) indicates that there is no effect, a question remains, at least for me,[38] as to what was the cause of the originally reported effect. The statistical analysis seems correct. Both DeRujula and I obtained the same slope that Fischbach found in our fits of the reanalyzed data. Perhaps a plausible explanation is that the uncertainties reported by Eötvös were underestimated. Scientists often do this (Franklin 1986, Chap. 8). I also note that using the original Eötvös data gives only a two standard deviation effect (Fig. 1.11). As Fischbach pointed out, however, Eötvös' method of analysis increased the uncertainties. It remains a puzzle. Nevertheless, I believe that it was reasonable for physicists to regard these three pieces of evidence as increasing the plausibility of the Fifth Force. At the time they were tantalizing, if uncertain, results.

---

[38]Both Fischbach and Newman have also raised this question with me. I suspect others have also asked it.

References

Ackermann, R.: Allan Franklin, right or wrong. In: Fine, A., Forbes, M., Wessels, L. (eds.), pp. 451–457 (1991)

Anderson, P.: The reverend Thomas Bayes, needles in haystacks, and the Fifth Force. Phys. Today **45**, 9–11 (1992)

Bell, J.S., Perring, J.: 2π decay of the K_2^0 meson. Phys. Rev. Lett. **13**, 348–349 (1964)

Bernstein, J.: A Comprehensible World. Random House, New York (1967)

Earman, J.: Bayes or Bust. MIT, Cambridge (1992)

Fine A., Forbes, M., Wessels, L. (eds.): PSA 1990, vol. 2. Philosophy of Science Association, East Lansing, MI (1991)

Fischbach, E., et al.: Reanalysis of the Eötvös experiment. Phys. Rev. Lett. **56**, 3–6 (1986)

Franklin, A.: The Neglect of Experiment. Cambridge University Press, Cambridge (1986)

Franklin, A.: Experiment, Right or Wrong. Cambridge University Press, Cambridge (1990)

Franklin, A., Howson, C.: Why do scientists prefer to vary their experiments? Stud. Hist. Philos. Sci. **15**, 51–62 (1984)

Galison, P.: How Experiments End. University of Chicago, Chicago (1987)

Howson, C., Urbach, P.: Scientific Reasoning: The Bayesian Approach. Open Court, La Salle (1989)

Lynch, M.: Allan Franklin's transcendental physics. In: Fine, A., Forbes, M., Wessels, L. (eds.), pp. 471–485 (1991)

Nelson, A.: How could scientific facts be socially constructed? Stud. Hist. Philos. Sci. **25**, 535–547 (1994)

Pickering, A.: The hunting of the quark. Isis **72**, 216–236 (1981a)

Pickering, A.: The role of interests in high-energy physics: the choice between charm and color. In: Knorr, K.D., et al. (eds.) The Social Process of Scientific Investigation. Sociology of the Sciences, vol. IV, 1980, pp. 107–138. Reidel, Dordrecht (1981b)

Pickering, A.: Constructing Quarks. University of Chicago, Chicago (1984a)

Pickering, A.: Against putting the phenomena first: the discovery of the weak neutral current. Stud. Hist. Philos. Sci. **15**, 85–117 (1984b)

Pickering, A.: Reason enough? More on parity-violation experiments and electroweak gauge theory. In: Fine, A., Forbes, M., Wessels, L. (eds.), pp. 459–469 (1991)

Pinch, T.: Confronting Nature. Reidel, Dordrecht (1986)

Thieberger, P.: Thieberger replies. Phys. Rev. Lett. **62**, 810 (1989)

Chapter 4
Conclusion

This history of the Fifth Force has allowed us to see some of the roles that experimental evidence plays in the contexts of discovery, of pursuit, and of justification. We have seen the supportive and suggestive role of evidence that led, in part, to the proposal of the Fifth Force hypothesis. It was also evidence that helped to provide sufficient plausibility so that a segment of the physics community thought the hypothesis worth further experimental and theoretical investigation. Finally, it was evidence that decided the ultimate fate of the hypothesis. As we have seen, the decision that the Fifth Force does not exist was based on an overwhelming preponderance of evidence. We have also discussed some of the other considerations that enter into the context of pursuit.

There may be those who, in the light of this decision, would say that the entire episode was a waste of time, effort, and money, and that it hardly merits serious historical and philosophical consideration. They might say that Feynman and Glashow were correct, and that the hypothesis should not have received the attention it did, particularly in light of the fact that the evidence originally supporting the Fifth Force is now regarded as incorrect. They would be wrong. Leaving aside the issue of hindsight, we have seen that hypotheses thought to be unlikely or incorrect, even by eminent scientists, have turned out to be correct. Science, as Feynman himself pointed out, is not decided by authorities, but on valid experimental evidence. Wrong physics is not bad science. The most important thing is to do it in the "right way", and this, I believe, was done in this case. Scientists used the best evidence available at the time. This is not an unreasonable procedure. One of the ways in which science progresses is by investigating speculative hypotheses. If the Fifth Force hypothesis, like that of parity nonconservation, had turned out to be correct, no one would question its investigation. They would, instead, applaud the courage, intuition, and insight of both those who proposed it, and those who investigated it. In addition, the episode has not been without value for physics. More stringent limits have been set on possible violations of a fundamental law, and experimental and calculational techniques have been improved.

© Springer International Publishing Switzerland 2016
A. Franklin, E. Fischbach, *The Rise and Fall of the Fifth Force*,
DOI 10.1007/978-3-319-28412-5_4

The episode is also important for the historian and philosopher of science. As Alexandre Koyre noted (1978, p. 66):

> What is served, after all, by dwelling on error? Is not the important thing the successful outcome, the discovery, and not the difficult, winding paths that had to be followed, and on which there was always the possibility of going astray.... What matters from the point of view of posterity is the victory, the discovery, the invention. However, for the historian of scientific thought, at least for the historian-philosopher, failure and error,... can sometimes be as valuable as their successes. They can, perhaps, be even more so. They are, in fact, very instructive. They sometimes enable us to grasp and understand the hidden processes of their thinking.

It is in both the decision to reject, as well as to accept, experimental results or hypotheses that we see the methodology of science.

> This little pig built a spaceship,
> This little pig paid the bill;
> This little pig made isotopes,
> This little pig ate a pill;
> And this little pig did nothing at all,
> But he's just a little pig still.   Winsor (1958, #10)

References

Koyre, A.: Galileo Studies. Humanities Press, Atlantic Highlands (1978)
Winsor, F.: The Space Child's Mother Goose. Simon and Schuster, New York (1958)

Part II
Additional Material

Chapter 5
The Fifth Force Since 1991

At the 1990 Moriond workshop, attended by many of those working on the Fifth Force, Orrin Fackler stated: "The Fifth Force is dead." No one disagreed. At the time there was no evidence that such a force, as initially proposed, with a strength approximately 1 % that of the gravitational force and a range of about 100 m, existed. More formally, Eric Adelberger and other members of the Eöt-Wash group concluded (Adelberger et al. 1990, p. 3291):

> We have made a sensitive, systematic search for interactions mediated by ultra-low-mass scalar or vector bosons using two different detector dipoles and two different sources. We find absolutely no evidence for any new interactions ascribable to such particles. Our results break new ground over ranges from roughly 1 AU down to roughly 30 cm,[1] and are considerably more precise than any of those which claim evidence for 'new physics'.

They further stated:

> Considerable experimental progress has occurred in the four years since 1986 when Fischbach et al. proposed a 'fifth force' (Fischbach et al. 1986). New experimental techniques have been introduced, and sensitivities have increased dramatically. The situation regarding inverse-square law tests is now clear. No violations are observed in astronomical or laboratory experiments. The experimental situation in tests of the universality of free fall (composition dependence), summarized in [Table 2.5 of the present book], has also been greatly improved. With two exceptions, experiments show no evidence for a new macroscopic interaction. The most sensitive, in terms of differential acceleration resolution, in each category—von Eötvös experiments with Earth and laboratory sources, Galileo experiments, and floating-ball experiments—give null results. Although reports of positive effects by Thieberger and by Boynton et al. have not been retracted, these authors themselves do not claim evidence for new physics (Adelberger et al. 1991, p. 306).

---

[1] This range included the suggested range for the Fifth Force of approximately 100 m.

© Springer International Publishing Switzerland 2016
A. Franklin, E. Fischbach, *The Rise and Fall of the Fifth Force*,
DOI 10.1007/978-3-319-28412-5_5

Ephraim Fischbach and Carrick Talmadge, two of the proposers of the initial hypothesis remarked (Fischbach and Talmadge 1992, p. 214):

> No compelling evidence has yet emerged that would indicate the presence of a fifth force, although the anomalies reported in the original Eötvös experiment remain to be understood, as do those in the experiments of Thieberger and Boynton et al.[2] On the experimental side, efforts continue to set even more stringent limits on possible deviations from Newtonian gravity, motivated in part by the recognition that such experiments may be our most powerful tool in exploring physics at the Planck scale.

Despite these obituaries, work on the Fifth Force, both experimental and theoretical, has continued into the twenty-first century. This includes explicit tests of the hypothesis. Other work, on the universality of free fall, on possible violation of Newton's inverse square law of gravity, and on the weak equivalence principle in general relativity, also has relevance for the Fifth Force. These later papers, although relevant, do not always mention the Fifth Force explicitly or cite the initial paper of Fischbach and his collaborators. The Eöt-Wash collaboration stated (Su et al. 1994, p. 3614):

> The universality of free fall (UFF) asserts that a point test body, shielded from all known interactions except gravity, has an acceleration that depends only on its location. The UFF is closely related to the gravitational equivalence principle, which requires an exact equality between gravitational mass m_g and inertial mass m_i and therefore the universality of gravitational acceleration. Experimental tests of the UFF have two aspects—they can be viewed as tests of the equivalence principle or as probes for new interactions that violate the UFF.

The UFF test would also test for the Fifth Force. The paper of Su et al. quoted above, for example, set limits on possible violations of Newton's Law of Universal Gravitation, and on a possible Fifth Force, but did not cite the 1986 paper of Fischbach and collaborators.

In this essay I will concentrate on the experimental work that has relevance for the Fifth Force which has taken place since Part I of this work was written in 1991. This is not intended to be a complete history, but rather to give the flavor of the variety of experimental work done on the Fifth Force at the end of the twentieth century and the beginning of the twenty-first century. We will find that the Fifth Force is still dead.

5.1   The 1990s

One of the earliest of these later experiments was performed by a group in China (Yang et al. 1991). The experimenters measured the differences in the acceleration due to gravity at various distances from an empty oil reservoir caused by filling or

---

[2]The effects observed in both the original Eötvös experiment and in Thieberger's experiment are still unexplained. Boynton's initial results have been superseded by his later results.

Table 5.1 Results of Yang et al. (1991)

Distance from central axis of water cylinder (m)	Mean experimental value Δg_e and its standard $(10^{-5}\,\text{m/s}^2)$	Newtonian prediction Δg_N deviation $(10^{-5}\,\text{m/s}^2)$	$\Delta g_e/\Delta g_N$
10.00	0.424 ± 0.002	0.423	1.002 ± 0.005
20.00	0.273 ± 0.002	0.272	1.004 ± 0.007
30.00	0.146 ± 0.002	0.145	1.007 ± 0.014
40.00	0.075 ± 0.002	0.073	1.027 ± 0.027
50.00	0.040 ± 0.003	0.038	1.053 ± 0.079

emptying the reservoir with water.[3] The acceleration was measured with a LaCoste–Romberg gravimeter, the standard apparatus used in earlier tower experiments. The experimenters compared the measured differences in acceleration with those calculated from Newtonian gravity alone. Any difference would be attributed to the Fifth Force. Their results are shown in Table 5.1. No differences between the measured and calculated values are seen. The group concluded (Yang et al. 1991, p. 332):

> It is worth pointing out that a weak intermediate-range interaction of Yukawa form is not excluded by our data but the possible strength of such an interaction is highly constrained $|\alpha| < 0.002$. This is in agreement with the results of the WTVD [Eckhardt's group] and BREN [Lawrence Livermore group] tower gravity experiments.[4]

The experimental group noted that the readings of the LaCoste–Romberg gravimeter varied with time because of drift and tidal effects. They were able to subtract these effects by comparing their gravimeter with another identical gravimeter located 200 m from the reservoir. They also remarked that the local topography was flat, avoiding the terrain problems that Bartlett and Tew had pointed out caused difficulties in the analysis of earlier experiments. There was also the possibility that the pumping operations affected the measurements of the accelerations, and reported that the effect was "negligible compared with the standard deviations quoted in the table" (p. 331).

In early 1992 a different type of test of a possible Fifth Force was reported by a group at the University of Washington (Venema et al. 1992).[5] This experiment searched for an interaction of the form $\sigma \cdot r$, in two isotopes of mercury, ^{199}Hg and ^{201}Hg. Here σ is the spin operator for the nucleus of the mercury isotope and r is the vector pointing toward the center of the earth. The experimenters measured the spin precession frequencies of the two isotopes for two orientations of the magnetic

---

[3] This was similar to Bennett's experiment at the lock on the Snake River, discussed in Part I.

[4] These were discussed in Part I.

[5] The group included Blayne Heckel, one of the leaders of the Eöt-Wash group.

field relative to the Earth's gravitational field. They remarked that (p. 135):

> The $\sigma \cdot r$ interaction could also arise from a new interaction coupled to something other than mass as has been discussed extensively in connection with recent experiments to detect a fifth force.[6]

They noted that the question of such an interaction was open because most previous experiments had used unpolarized test bodies. They found that a spin-dependent component of gravitational energy was less than 2.2×10^{-21} eV. Their conclusion was (p. 135):

> Our result provides a test of the equivalence principle for nuclear spins, and sets limits on the magnitude of possible scalar–pseudoscalar interactions which would couple to spins.

There was still no evidence for a Fifth Force.

There were also replications of previous types of experiment. Liu et al. (1992) measured the acceleration due to gravity as a function of height on a 320 m tower. This would test the possible distance dependence of the Fifth Force. They noted the previous discord between the early positive results reported by Eckhardt and his collaborators and the negative results reported by the Lawrence Livermore group, by Speake et al., by the later results of Eckhardt's group, and by others. They remarked (p. 131):

> Many have questioned the results of Eckhardt et al. including Thomas et al. [the Livermore group] who, in an independent tower (BREN tower) experiment, found no evidence for non-Newtonian gravity. More recently Eckhardt et al. have revised their analysis and now their results appear consistent with Newtonian gravity. The newer and more precise Erie tower results of Cruz et al. (1991) now set a little stronger constraints on such a kind of non-Newtonian force. We decided that an independent experiment would help clarify the situation, and undertook to perform a tower test of gravity.

The experimenters used the standard LaCoste–Romberg gravimeter and corrected their results for tides, drift, gravimeter screw errors, and systematic effects due to tower motion. (All measurements were done at wind speeds less than 3 m/s.) They stated that their tower was stable and located on a nearly flat terrain. Their results are shown in Fig. 5.1, along with both the old and new results of Eckhardt et al. and several of the newer results. They concluded (p. 131):

> In a tower test of Newton's inverse square law of gravitation we found no evidence for the non-Newtonian force, and the accuracy of the experiments constrains the Yukawa potential coupling constant $|\alpha|$ to be less than 0.0005.

Carusotto and collaborators (1992) performed an interesting variant on the Galileo-type free fall experiments discussed earlier. The experiment was performed at the surface of the Earth in a vacuum chamber. They measured the angular

---

[6]Recall the earlier discussions of possible coupling to either baryon or lepton number.

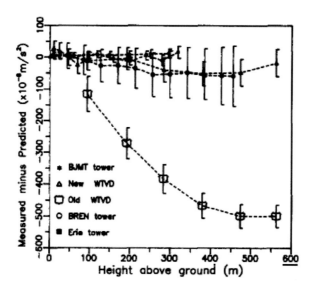

Fig. 5.1 Measured minus predicted values of the acceleration due to gravity as a function of the height above ground for various tower experiments (From Liu et al. 1992)

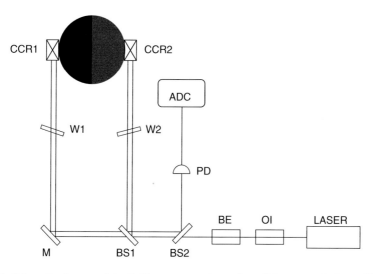

Fig. 5.2 Schematic diagram of the Galileo-type experiment for a disk composed of two different metals (From Carusotto 1993)

acceleration of a disk which had a half-disk of aluminum and a half-disk of copper (Fig. 5.2) (Carusotto et al. 1992, p. 1723):

> If there is a difference Δg in the free-fall acceleration of aluminum and copper, then the disk assembly experiences a torque and, therefore there is an angular acceleration of the disk assembly.

The disk would rotate. The acceleration was measured using laser light reflected from corner reflectors placed on the disk. The experimenters checked the sensitivity of their apparatus and looked for possible systematic effects by first making measurements with a disk made only of aluminum. They found $\Delta g/g = (3.2 \pm 9.5) \times 10^{-10}$, consistent with zero. There were no large systematic effects. Using the half-copper half-aluminum disk they found $\Delta g/g = (8.5 \pm 9.5) \times 10^{-10}$ and $\Delta g/g = (-4.8 \pm 11.2) \times 10^{-10}$ with the disk reversed. They combined the two sets of measurements and set a limit of $\Delta g/g = (2.9 \pm 7.2) \times 10^{-10}$ (p. 1725):

> The result is compatible with zero (no g violation) and it is in quite good agreement with the one obtained by Kuroda and Mio for the same materials.

At this same time three other innovative and difficult experiments were proposed that would search for the Fifth Force and also test the law of gravity, viz., Pace et al. (1992), Sanders and Deeds (1992), and Slobodrian (1992). These proposed experiments would be conducted either in space or in free fall from a large height (40 km). As Sanders and Deeds remarked (1992, p. 489):

> Much of the difficulty in gravitational measurements arises from the extreme weakness of the gravitational force between the test bodies compared to other forces acting on the bodies such as electromagnetic effects and instrumental friction. Space is attractive for gravitation measurement because it has the potential to be relatively 'clean' and free of the influences which necessarily cloud the interpretation of terrestrial experiments.

Slobodrian agreed (1992):

> All experiments to date have been carried out near the Earth's surface and the test bodies are subject to the strong force due to the Earth's attraction.

None of these experiments was ever performed, but they indicate the significant amount of interest in the Fifth Force, even after its presumed demise. Only one of these proposed experiments, that of Sanders and Deeds, attracted any attention in the physics literature. The citations to the paper of Sanders and Deeds were comments on the general method proposed, suggestions of other similar experiments, and discussions of possible problems in the experiment.

Although a discussion of these papers is somewhat peripheral to our history of the Fifth Force, the physics involved in the experiments is fascinating and the papers also address a problem current both then and now, the measurement of G, the universal gravitational constant. I begin with the experiment proposed by Sanders and Deeds. The purpose of the experiment was threefold. As the authors remarked (Sanders and Deeds 1992, p. 489):

> The first 'constant of nature' to be identified, Newton's constant of universal gravitation, G, is presently the least accurately known. The currently accepted value is $(6.67259 \pm 0.00085) \times 10^{-11}$ m^3 kg^{-1} s^{-2} has an uncertainty of 128 parts per million (ppm),[7] whereas most other fundamental constants are known to less than 1 ppm. Moreover, the inverse-square law and the equivalence principle are not well validated at distances of the order of meters. We propose measurements within an orbiting satellite which would improve

---

[7]The current uncertainty on the value of G is 120 ppm. The latest CODATA (Committee on Data for Science and Technology) value is $(6.673\,84 \pm 0.000\,80) \times 10^{-11}$ m^3 kg^{-1} s^{-2}.

the accuracy of G by two orders of magnitude and also place upper limits on the field-strength parameter α of any Yukawa-type force, assuming a null result. Preliminary analysis indicates that a test of the time variation of G may also be possible.

The method proposed was as follows (pp. 489–490):

> The satellite energy exchange (SEE) method would measure the gravitational interaction between two test bodies by placing them in nearly identical Earth orbits and treating their interaction by orbital perturbation techniques, which historically have enjoyed unsurpassed accuracy.

This problem had been studied as early as 1897 by George Darwin. He had found that for two satellites in identical orbits their mutual gravitational interaction could result in 'horseshoe' orbits. The interaction would slow down the leading satellite, which would then move to a lower orbit, whereas the trailing satellite would accelerate and move to a higher orbit. This would result, if initial conditions were right, in the horseshoe orbits shown in Fig. 5.3.[8] This had been used by Dermott and Murray (1981) to explain apparent anomalies in the orbits of two satellites of Saturn. Sanders and Deeds noted that encounters between two such satellites would not necessarily result in horseshoe orbits, but that the experiment could still be performed (Sanders and Deeds 1992, p. 491):

> The interaction between the two test bodies in a SEE [Satellite Energy Exchange] will test for both violations of the inverse-square law and for composition-dependent difference (Eötvös' experiment), which may also be interpreted as violations of the equivalence principle.

It would also measure G at a distance of meters. The proposed experiment is shown in Fig. 5.4. It includes a large 'shepherd' mass and a smaller mass (p. 495):

Fig. 5.3 Partly schematic diagram of the orbital configuration of the co-orbital satellites of Saturn, 1980S1 and 1980S3 (After Dermott and Murray 1981). Note that the satellites recede from each other after each encounter (From Sanders and Deeds 1992)

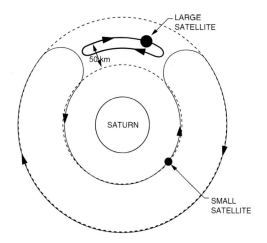

---

[8]This appears to show an apparently repulsive gravitational force, a rather odd result. The force is, of course, attractive.

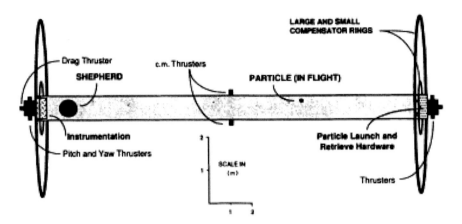

Fig. 5.4 Diagram of the satellite experiment proposed by Sanders and Deeds (1992)

The interacting masses need to be enclosed inside a conducting shell, called the 'capsule', to protect them from atmospheric drag, radiation pressure from both the Sun and the Earth, and electric and magnetic fields, all of which would be at least comparable to the gravitational interaction between the masses. The capsule would be equipped with a system of optical lasers and interferometric sensors to monitor its own size and shape and the positions of the shepherd and particle.

Analysis of the observed motions could be used to obtain the needed results. Extensive calculations of experimental uncertainty indicated that $\Delta G/G$ could be measured to $\sim 1 \times 10^{-6}$ and would set a limit on α, the strength of the Fifth Force, $|\alpha| < 1.2 \times 10^{-6}$ for the most sensitive distance between the masses of the order of a few meters.

The experiment proposed by Slobodrian (1992) made use of the idea that if a small test object were dropped into a hole drilled through the center of an assumed homogeneous Earth, then the motion of the test object would be simple harmonic motion. He proposed drilling a cylindrical hole through the center of a large copper sphere and then dropping a small test object of either copper or beryllium into the hole. The period of that motion would be observed using a hole drilled perpendicular to the cylinder, this would allow the calculation of G_{ij} the gravitational constant between the two materials along with α, the Fifth Force strength. The small effects of the cylinder and of the hole could be easily calculated numerically. The experiment would be conducted in, but not attached to, the Space Station to minimize the effect of the Earth's attraction. For an assumed precision of 10^{-4} s in the time measurement of the period of the small test object, and a period he had calculated of between 48 and 145 min for test object of different substances, a precision in $\Delta \tau/\tau$ of the order of 3×10^{-8} would be obtained. This would improve the precision in the measurements of G and α by about a factor of 700.

Another novel experiment, was proposed by Pace et al. (1992). The experiment proposed to measure the differential acceleration due to gravity for two bodies of different composition, falling from a height of 40 km. The authors described their experiment as follows (Pace et al. 1992, p. 3112):

> A special system of free falling objects has been developed to perform this experiment. A brass cylinder horizontally oriented is inserted in another cylinder of stainless steel; four capacitive transducers (CTs) are applied to this system to detect the gap variations between the surfaces of the cylinders. Each transducer is an element of a microtransmitter and its capacitance determines the value of the resonance frequency: any variation of distance between the surfaces of the cylinders produces a variation of the resonance frequency. In this way we may detect distance variations of about 0.1 μm, so that the sensitivity is $\Delta a/g \sim 10^{-11}$, corresponding to a differential acceleration $\Delta a = 1.38 \times 10^{-10}$ ms^{-2}.

This was, however, larger than the value $\Delta a = (4.5 \pm 4.4) \times 10^{-13}$ ms^{-2} already obtained for the differential acceleration by the Eöt-Wash group in 1989 for their beryllium–aluminum pendulum.

Experimental tests of the Fifth Force hypothesis continued in 1993. The group at the Tata Institute, using a torsion pendulum, set more stringent limits on the possible coupling to isospin. Their 2σ limit for the strength was $-5.9 \times 10^{-5} \leq \alpha_1 \leq 3.44 \times 10^{-5}$, "the best upper limit on α_1 for all the experiments so far" (Unnikrishnan 1993, p. 408). Carusotto and collaborators reported further results on their falling-disk experiment (Table 5.2). In this experiment they used a copper–tungsten disk, rather than a copper–aluminum disk. They concluded (Carusotto 1993, p. 357):

> There is no evidence for any g-universality violation, at the level of μGal, at least with the Galileo-type experiment performed so far.

In mid-1993 a symposium on experimental gravitation was held in Nathiagali, Pakistan. Invited papers from the symposium were published in a special issue of Classical and Quantum Gravity (Volume 6A, 1994). Several of the papers were relevant to tests of the Fifth Force. Eric Adelberger presented an early version of the results later published as Su et al. (1994) and Unnikrishnan presented a review of on experimental gravitation in India, which included the results he published in (Unnikrishnan 1993). Another group at the University of Washington, not the Eöt-Wash collaboration, proposed a new class of experiments to test the inverse-square law of gravity. These experiments would use a torsion pendulum (Moore et al. 1994, p. A97):

> [...] whose mass distribution is specifically configured to provide high-sensitivity detection of a uniquely non-Newtonian derivative of the potential (the horizontal derivative of the Laplacian), rather than looking for a small deviation from the expected power-law

Table 5.2 Results of Carusotto (1993)

References	Compared materials	Δg (μGal)
Present work	Cu–W	0.71 ± 0.91
Carusotto et al. (1992)	Al–Cu	0.29 ± 0.72
	Al–Cu	-0.13 ± 0.78
	Al–Be	0.43 ± 1.23
Kuroda and Mio (1990)	Al–C	-0.18 ± 1.38
Niebauer et al. (1987)	Cu–U	0.13 ± 0.5

dependence on distance of a Newtonian field derivative. This method provides a stronger null test of the gravitational inverse-square law force because it is less sensitive to imperfections in the source mass. We discuss the design of these experiments and estimate their performance relative to currently established experimental limits on inverse-square law violation.[9]

In 1994 Eckhardt's group[10] published results on measurements of the acceleration due to gravity as a function of the height of the measurement on a tower, using a tower different from the one they had used in their previous experiments (Romaides et al. 1994). They noted that they had initially obtained results at the WTVD tower in North Carolina which showed an apparent violation of Newtonian gravitation, but that their later results, along with those of other tower experiments had shown that Newton's law of gravity was valid over a range from 10 m to 10 km. (This was discussed in Part I.) They stated that (Romaides et al. 1994, p. 3608):

> Two of the major difficulties in the experiment were the inaccessibility of some areas around the WTVD tower, and the lack of a good terrain model, which meant that some computations could not be done as rigorously as desired.

Their new results were obtained at the WABG tower in Mississippi, which had the advantage of very flat local terrain and easy access for gravity measurements near the tower. They remarked that they had been unable to obtain measurements at the largest height of 571 m, an omission they would later remedy. They concluded (p. 3608):

> The tower observations were compared to the predictions, with the largest discrepancy being $-33 \pm 30 \,\mu$Gal at 493 m. The results are in good agreement with previous tower experiments, which also are in accord with the inverse-square law, and they set further restrictions on possible non-Newtonian forces.

The group reported that their WABG results agreed not only with their last WTVD tower results but also with the results of other tower experiments (Fig. 5.5). They stated that they were ending their investigations[11] and that (p. 3612):

> [...] we have learned from these and other experiments that there is no credible evidence for deviations from the inverse-square law over a laboratory to solar system scale length. By helping to fill in the scale $\lambda \approx 10^3$ m, tower experiments have thus played an important role in confirming our belief in the validity of Newtonian gravity.

The inclusion of tests of the Fifth Force as part of more general experimental work on general relativity and its implications became clear in the 1994 report of the Eöt-Wash group mentioned earlier (Su et al. 1994). The experimenters stated purpose was to measure the universality of free fall with respect to the Earth, the

---

[9]The group included Paul Boynton, who had reported both positive and negative earlier results on the Fifth Force. Boynton's later negative results were regarded as superseding his earlier work.

[10]The group also included Fischbach and Talmadge, two of the initial proposers of the Fifth Force hypothesis.

[11]As we shall see below, this is not quite accurate.

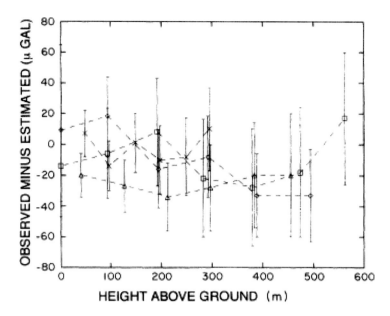

Fig. 5.5 The observed-minus-model discrepancies for all tower experiments along with their associated errors. *Diamonds* are the WABG results, *boxes* are the WTVD results, *triangles* are the BREN tower results, and *crosses* are the Erie tower results. In order to avoid clutter, not all data points were plotted. Note the excellent agreement especially at the upper elevations (From Romaides et al. 1994)

Sun, our galaxy, and in the direction of the cosmic microwave dipole.[12] They further noted that (p. 3614):

> Our galactic-source results tests the UFF [Universality of free fall] for ordinary matter attracted toward dark matter.[13]

The experimental group had made improvements in their torsion balance apparatus including better regulation of the turntable speed, compensation for gravity gradients, and in the calibration of their instruments. Although the Fifth Force is not explicitly mentioned, nor is the paper of Fischbach et al. cited, the Eöt-Wash results did provide more stringent limits on the presence of such a force. It is difficult to make a direct comparison between the earlier and later results because the 1991 Eöt-Wash paper presented a limit on a force with a range of 30 m, whereas their 1994

---

[12]The title of the paper was: "New tests of the universality of free fall."

[13]The group also stated that (Su et al. 1994, p. 3614):

> We also test Weber's claim that solar neutrinos scatter coherently from single crystals with cross-sections $\sim 10^{23}$ times larger than the generally accepted value and rule out the existence of such cross-sections.

For a more detailed history of this episode see Franklin (2010).

Table 5.3 Comparison of the 1991 (Adelberger et al. 1991) and 1994 (Su et al. 1994) Eöt-Wash results for $\alpha\Delta(B/\mu)_{\text{detector}}(B/\mu)_{\text{source}}$

	$\lambda = 30\,\text{m}$	$\lambda = 20\,\text{m}$	$\lambda = 50\,\text{m}$
1991	$(1.4 \pm 2.9) \times 10^{-8}$		
1991	$(-2.1 \pm 3.6) \times 10^{-8}$		
1994 (Be–Al detector)		$(-0.5 \pm 1.1) \times 10^{-8}$	$(-2.6 \pm 5.4) \times 10^{-9}$
1994 (Be–Cu detector)		$(-11 \pm 9.8) \times 10^{-9}$	$(-5.3 \pm 4.8) \times 10^{-9}$

Fig. 5.6 Schematic view of the Gigerwald experiment (From Cornaz et al. 1994)

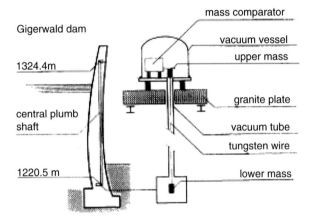

paper gave limits for both 20 and 50 m. The results are shown in Table 5.3. One can see that the uncertainty in the results has improved by a factor of approximately three.

A group at the University of Zurich reported another test of the Fifth Force (Cornaz et al. 1994).[14] The experiment measured the difference in weight between two masses as a function of the height of the water in a pumped-storage reservoir, Lake Gigerwald (Fig. 5.6) (p. 1152):

> The basic idea of the Gigerwald experiment was to measure the weight difference of two test masses located above and below the variable water level with a single balance.

The experimental design avoided several of the problems of such experiments (pp. 1152–1153):

> Since the weight difference is measured in a short time, balance drifts are negligible. Time-variable gravity effects originating from distances much larger than the separation of test masses completely vanish (e.g., tides). By comparing the weight differences at several water levels even the static local gravity from the surroundings cancels. Finally, the recorded gravity signal is just due to the interaction between the locally moved mass (water and air) and the test masses.

---

[14]The major purpose of the experiment, as the title of the paper reveals, was to measure G, the gravitational constant.

Fig. 5.7 The *solid curve* is the calculated weight difference of the two test masses as a function of the water level following pure Newtonian gravity (the origin is set at 1240 m for an empty lake) (From Cornaz et al. 1994)

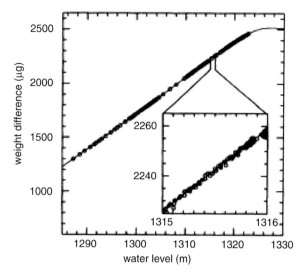

Fig. 5.8 Excluded strengths α, and ranges λ for a single Yukawa model at the 2σ level arising from experiments measuring directly the gravitational constant at geophysical distances (From Cornaz et al. 1994)

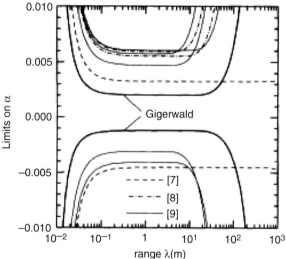

The comparison between the theoretically calculated weight differences and the measured values is shown in Fig. 5.7. The experimenters obtained more stringent limits on α, the strength of the proposed force, as a function of λ, the range, than had been obtained in previous experiments (Fig. 5.8).[15]

Experimental work on tests of the Fifth Force slowed, although there was still considerable theoretical work. In 1996, Carusotto et al. published their final results,

---

[15]This experiment was similar to those of Moore et al. (1988) and Bennett (1989).

which were the same as those discussed earlier, except for the inclusion of a small systematic uncertainty. They concluded (Carusotto et al. 1996, p. 1274):

> There is no evidence of any g-universality violation, at the level of μGal, at least with the Galileo-type experiment performed so far.

In 1997, Romaides et al. published their final results from the WABG tower experiment. They had overcome the difficulties in making measurements at the largest height and stated (Romaides et al. 1997, p. 4532):

> [...] we succeeded in obtaining readings at 568 m above ground level. These readings, along with the previous results on the WABG and WTVD towers, allow for even tighter constraints on the non-Newtonian force parameters α and λ [the strength and range of the proposed Fifth Force]. Furthermore, we can now combine our tower data with data from lake experiments to give very tight constraints on the non-Newtonian coupling constant α over the entire geophysical window (10 m to 10 km).

Those constraints are shown in Fig. 5.9. They concluded (p. 4536):

> In summary, we conclude from existing tower experiments that at the present time there is no evidence for any significant deviation from the inverse-square law for $\lambda \approx 10^3$ m.

The Eöt-Wash group reported a new result using an interesting variant on their previous experimental apparatus (Gundlach et al. 1997). In their previous work the group had used a torsion balance mounted on a rotating platform to measure the differential acceleration of various substances toward a local hillside, and to other sources such as the Sun, the Earth, and the galaxy. In their latest experiment the

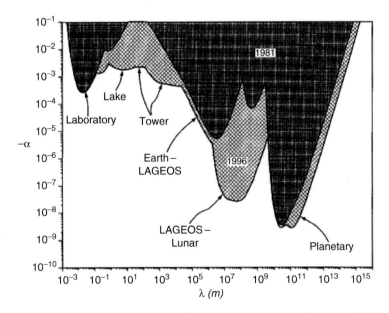

Fig. 5.9 The constraints on α as a function of λ in 1981 (*dark region*) and again in 1996 (*hatched region*) after including the most recent experimental results (From Romaides et al. 1997)

Fig. 5.10 Schematic view of the Rot-Wash instrument. The ^{238}U was counterbalanced by 820 kg of lead so the floor would not tilt as the attractor revolved (From Gundlach et al. 1997)

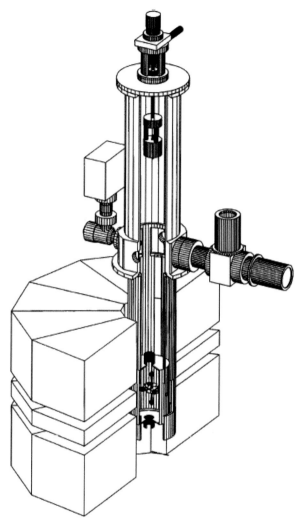

experimenters used a rotating three-ton ^{238}U attractor to measure the differential acceleration of lead and copper masses placed on a torsion balance. The Rot-Wash[16] apparatus is shown in Fig. 5.10. The surroundings of the torsion balance were temperature controlled to guard against possible temperature effects. The ^{238}U was counterbalanced by 820 kg of Pb so the floor would not tilt as the attractor revolved. As discussed earlier, tilt was a significant source of possible background effects in the Eöt-Wash experiments. The reason for the modification of the apparatus was that their previous experiment (Su et al. 1994) had been unable to test for forces with a

---

[16]The Eöt-Wash group continued its whimsy with the naming of their new apparatus.

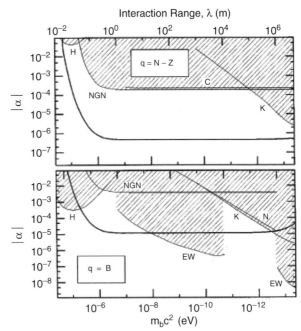

Fig. 5.11 2σ constraints on $|\alpha|$, the strength of the interaction, as a function of λ, the range of the interaction. The *heavy curves* are the Rot-Wash results and the *shaded region* shows previous results (From Gundlach et al. 1997)

range from 10 to 1000 km. The new apparatus, using a local source, allowed such a test. The experimenters concluded (Gundlach et al. 1997, p. 2523):

> We found that $a_{Cu} - a_{Pb} = (-0.7 \pm 5.7) \times 10^{-13}$ cm/s^2, compared to the 9.8×10^{-5} cm/s^2 gravitational acceleration toward the attractor. Our results set new constraints on equivalence-principle violating interactions with Yukawa ranges down to 1 cm and rule out an earlier suggestion of a Yukawa interaction coupled predominantly to $N - Z$.

The group stated that the new results improved the limits on the presence of forces proportional to $N - Z$ (number of neutrons minus the number of protons) and proportional to baryon number (B) by a factor of approximately 300 (Fig. 5.11).

In 1997 George Gillies published a review of measurements of the gravitational constant and other related measurements. He remarked that (Gillies 1997, p. 200):

> The contemporaneous suggestion by Fischbach et al. (1986) that there may be previously undiscovered, weak, long-range forces in nature provided further impetus for investigating the composition- and distance-dependence of gravity, since the presence of any such effect might reveal the existence of a new force. During this time, a theoretical framework for admitting non-Newtonian effects into discussions of the experimental results was emerging. It led to the practice of using the laboratory data to set limits on the size of the strength-range parameters in a Yukawa term added onto the Newtonian potential, and this has become a standard method for intercomparing the results of this class of experiments. Even though convincing evidence in favour of such new weak forces was never found, the many resulting experiments, when viewed as tests of the universality of free-fall, did much to improve the experimental underpinnings of the weak equivalence principle (WEP) of general relativity. In fact, searches for departures from the inverse square behaviour of Newtonian gravity have now come to be interpreted as attempts to uncover violations of the WEP.

Fig. 5.12 Sketch of the Lake Brasimone experiment (From Achilli et al. 1997)

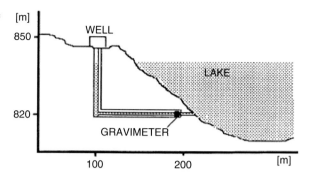

Fig. 5.13 Schematic cross-sectional view of the gravity sensor. The entire apparatus is contained in a liquid helium bath (From Achilli et al. 1997)

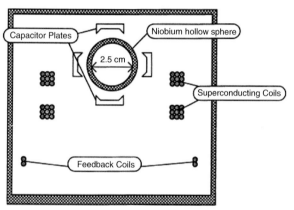

After a decade of negative experimental results of the Fifth Force, 1997 produced a positive result. Achilli et al. (1997), using a super-conducting gravimeter, measured changes in the gravitational force caused by the changing water level in a pumped storage reservoir, Lake Brasimone in Italy, found evidence for a violation in the distance dependence of Newton's law (Fig. 5.12). The superconducting gravimeter could measure variations in gravity of the order of 1 nGal (1 Gal $=$ 1 cm/s^2). A problem for the experimenters was the fact that tidal effects were of the order of 100–250 μGal. That effect could not be calculated precisely so the group measured the lake tides for a period of five months at a location 400 m from the lake. The experimenters also obtained a detailed survey of the lake shore, an important factor in obtaining a result.

The gravimeter measured the gravitation effect by measuring the feedback force needed to maintain a levitated superconducting niobium sphere in a fixed position (Fig. 5.13). They calibrated their apparatus by moving a known annular mass vertically with the gravimeter at its center. They also compared their gravimeter to an absolute gravimeter from another laboratory. The experimenters also investigated and measured geological, temperature, water table, and density background effects. Their final result $R =$ Observed/Theoretical effect was 1.0127 ± 0.0013 (Achilli et al. 1997, p. 775):

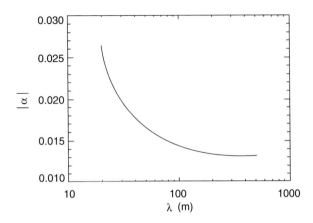

The ratio between the measured and expected gravitational effects differs from 1 by more than 9 standard deviations.

The experimenters noted, however, that (p. 802):

[...] the only parameter not verified at the 0.1% level was the gravimeter calibration factor. In any case, the adopted value is in agreement with the result of the comparison with an absolute gravimeter.

Their results for $|\alpha|$ as a function of λ are shown in Fig. 5.14. The group stated that their result differed from that found by Cornaz et al. in a similar experiment (see earlier discussion). They remarked that a possible explanation for the discrepancy was the effective interaction distances of the water masses in the two experiments. Their $r_{\text{eff}} = 47$ m, whereas that of Cormaz et al. was $r_{\text{eff}} = 112$ m. In some theories this was an important difference.

5.2   The Twenty-First Century

The Eöt-Wash group continued taking data with their rotating ^{238}U attractor. They remarked that (Smith et al. 2000, p. 022001–1):

Our new results set new constraints on equivalence principle violating interactions with Yukawa ranges down to 1 cm, and improved by substantial factors existing limit for ranges between 10 km and 1000 km.

Their new value for the difference in acceleration for copper and lead masses was $a_{\text{Cu}} - a_{\text{Pb}} = (-1.0 \pm 2.8) \times 10^{-13}$ cm/s^2, with the uncertainty reduced by a factor of two compared to their 1997 result. Their results are shown in Fig. 5.15.

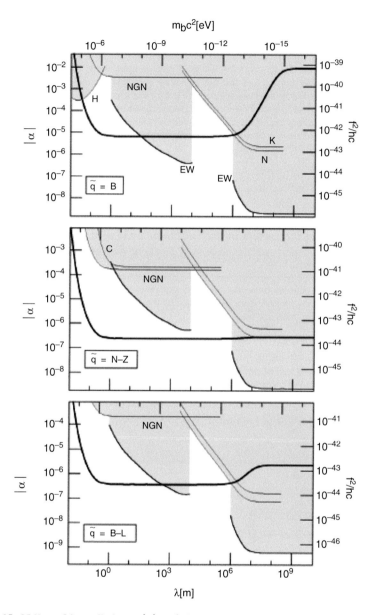

Fig. 5.15 95 % confidence limits on $|\alpha|$ vs λ for hypothetical interactions coupling to vector charges $q = B$, $q = N - Z$, or $q = B - L$, where B is baryon number, N is the number of neutrons in the nucleus, Z is the number of protons, and L is the number of leptons. The *heavy curves* are from this work (From Smith et al. 2000)

Other possible sources of variations of the gravitational force were also investigated by the Eöt-Wash group. They noted that (Heckel et al. 2000b, p. 153):

> The extraordinary sensitivity of the torsion balance has made it a valuable tool to test symmetries in nature and to search for new weak macroscopic forces. Most torsion balance experiments employ unpolarized test bodies either of different composition to test the universality of free fall or of special geometry to test for violations of the $1/r^2$ law of gravity. There are several motivations, however, to perform similar torsion balance measurements with spin polarized test bodies: to help elucidate the role of spin in gravitation, to search for new forces mediated by pseudoscalar bosons, and to perform a precise test of Lorentz (rotational) and CPT invariance.

Their octagonal spin pendulum was composed of four Alnico magnets and four SmCo magnets. This resulted in a net spin polarization of the pendulum. This pendulum was then mounted within the Eöt-Wash II torsion balance apparatus.[17] This included extensive magnetic shielding. The effectiveness of the shielding was tested by reversing the current in the Helmholtz coils used to cancel the Earth's magnetic field. This increased the laboratory magnetic field by a factor of 50 (Heckel et al. 2000a, p. 1230):

> No statistically significant torque was observed with the enhanced magnetic field, giving us confidence that with the laboratory fields nulled by the Helmholtz coils, magnetic torques were well below the statistical noise level.

The limits on the possible spin-dependent corrections to the gravitational force are shown in Fig. 5.16, along with those of previous experiments.

Another proposed test of the universality of free fall, and also of the composition-dependence of the Fifth Force was described by Nobili et al. (2000). In this experiment two concentric spinning masses composed of different substances would be contained in an orbiting satellite (Fig. 5.17). Any difference in the gravitational

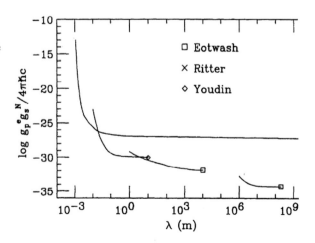

Fig. 5.16 2σ constraints on the spin coupling constants versus λ. The Eöt-wash curve is the result reported here (From Heckel et al. 2000a)

---

[17]This was an improved version of the original Eöt-Wash torsion balance.

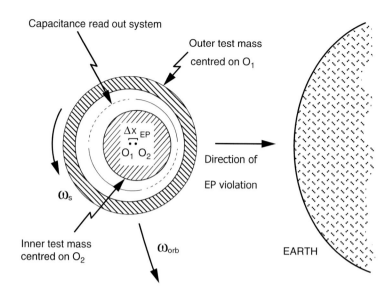

Fig. 5.17 A sketch of the proposed Galileo-Galilei equivalence principle test experiment (From Nobili et al. 2000)

force on the two masses would result in a spatial separation and detected by capacitance sensors. This was a space version of the experiment proposed earlier by Pace et al. (1992), which used falling objects.

Perhaps the most interesting result reported in 2000 was the withdrawal of the positive Fifth Force result of Achilli et al. (1997). As Focardi, a member of the group remarked (Focardi 2002, p. 419):

> The above result [the positive result] convinced us of the importance of making any possible effort to check the conclusions reached in the previous experiment.[18]

This withdrawal was based on a reanalysis of the same data used in the 1997 paper. (A more detailed discussion of the reanalysis appeared in Baldi et al. 2001.) The experimenters performed a new and better calibration of their superconducting gravimeter and included a more consistent model of tidal gravity variations. Recall that their initial paper had stated that (Achilli et al. 1997, p. 802):

> [...] the only parameter not verified at the 0.1% level was the gravimeter calibration factor.

Their new result for

$$R = \text{experimental value/theoretical calculation} = 1.0023 \pm 0.0017 \, .$$

---

[18]Focardi's paper was presented at a conference in 2000, but the conference proceedings were not published until 2002.

This should be compared with their earlier result of $R = 1.023 \pm 0.0017$. They concluded that (Baldi et al. 2001, p. 082001–2):

> The result of this analysis shows an agreement between data and Newtonian theory to within 0.1% level.

At the turn of the twenty-first century there was still no evidence supporting the Fifth Force.

In 2001 Bennett reported a second result from his experiment conducted at the Little Goose Lock on the Snake River. This was a torsion pendulum experiment which used the changing amount of water in the lock as an attractor. His initial data was taken in 1988 and published in 1989 (Bennett 1989). His 2001 paper (Bennett 2001) included additional data taken in 1990.[19] Bennett had made improvements in his apparatus including replacing the copper–lead disk in his torsion pendulum with a copper–lead annular ring (Bennett 2001, p. 123):

> A 2σ limit was set on the 'isospin coupling constant' of $\alpha_0 = \pm 0.001$ at $\lambda = 100$ m.

He also presented a summary of the 1σ limits on the differential acceleration for various pairs of substances (Table 5.4) along with a comparison of the coupling constants, α_0, obtained by various experiments (Fig. 5.18). The Fifth Force was still absent.

Despite the negative evidence, new experimental tests of the Fifth Force and of the weak equivalence principle were still being planned. Dittus and Mehls (2001), for example, were building a free-fall experiment in which two test masses of different substances would be dropped from a height of 110 m at the Bremen Tower. Any difference in fall would be detected by a SQUID (superconducting quantum interference device). They were aiming at an accuracy of better than 10^{-12} in the Eötvös ratio

$$\eta = 2 \frac{(m_\mathrm{g}/m_\mathrm{i})_1 - (m_\mathrm{g}/m_\mathrm{i})_2}{(m_\mathrm{g}/m_\mathrm{i})_1 + (m_\mathrm{g}/m_\mathrm{i})_2},$$

Table 5.4 Comparison of 1σ limits on differential acceleration (From Bennett 2001)

Reference	$\Delta a \times 10^{10}$ cm/s^2	Test masses	Source
Thieberger (1987c)	850 ± 260	Cu–H$_2$O	Cliff
Fitch et al. (1998)	30 ± 49	Cu–CH$_2$	Sloping terrain
Bennett (1989)	25 ± 52	Cu–Pb	H$_2$0
Bennett (2001)	2 ± 22	Cu–Pb	H$_2$0
Adelberger et al. (1990)	-0.15 ± 2.6	Be–Al	Pb

---

[19]For various personal reasons Bennett did not publish these results until 2001.

Fig. 5.18 Comparison of
different determinations of
the intrinsic coupling
coefficient α_0 for isospin
coupling (Note that Bennett
(1990) in the figure is Bennett
(2001) from which the figure
has been taken)

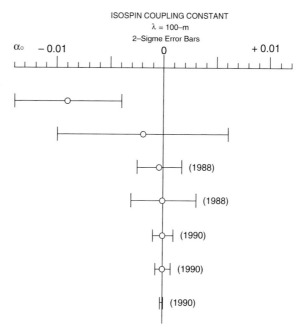

where m_i and m_g are the inertial and gravitational masses and the indices 1 and 2 are
for the test masses of different substances. They remarked that the then current best
value for η was less than 10^{-12} obtained by the Eöt-Wash group (Su et al. 1994).

Reasenberg and Phillips were developing a different type of apparatus (Reasenberg and Phillips 2001, p. 2435):

> We are developing a Galilean test of the equivalence principle in which two pairs of test
> mass assemblies (TMA) are in free fall in a comoving vacuum chamber for about 0.9 s. The
> TMA are tossed upward, and the process repeats at 1.2 s intervals.[20] Each TMA carries a
> solid quartz retroreflector and a payload mass of about one-third of the total TMA mass. The
> relative vertical motion of the TMA of each pair is monitored by a laser gauge working in an
> optical cavity formed by the retroreflectors. Single-toss precision of the relative acceleration
> of a single pair of TMA is 3.5×10^{-12} g. The project goal of $\Delta g/g = 10^{-13}$ can be reached
> in a single night's run.

In 2002 as part of a proposed satellite experiment to test the weak equivalence
principle, Moffat and Gillies summarized the current state of such tests (Moffat and
Gillies 2002, p. 92.3):

> In a long series of elegant experiments with rotating torsion balances, the Eöt-Wash
> group has searched for composition dependence in the gravitational force via tests of the
> universality of free fall. In terms of the standard Eötvös parameter η, they have reached
> sensitivities of $\eta \sim 1.1 \times 10^{-12}$ in comparisons of the accelerations of Be and Al/Cu test
> masses and, more recently, have resolved differential accelerations of approximately $1.0 \times$

---

[20]The title of their paper is: "Testing the equivalence principle on a trampoline."

10^{-14} cm s^{-2} in experiments with other masses. Drop-tower experiments now underway in Germany have as their goal testing WEP at sensitivities of $\eta \sim 1 \times 10^{-13}$, and Unnikrishnan describes a methodology under study at the Tata Institute of Fundamental Research in India wherein torsion balance experiments aiming at sensitivities of $\eta \sim 1 \times 10^{-14}$ are being developed.

None of these experiments provided evidence for the Fifth Force. The authors noted that proposed space-based experiments expected greater sensitivity. It was not clear, however, whether such experiments would cast any light on the Fifth Force, as initially proposed.

There were no other significant experimental tests of the Fifth Force in the early part of the twenty-first century. There were, however, experiments to measure G, the universal gravitational constant, a parameter whose value was then, and is now, uncertain. There were also experiments testing the law of gravity at very short distances, as well as continued discussions of space experiments. In 2005, Jens Gundlach, a member of the Eöt-Wash collaboration, published a review of the evidence to that date. His conclusion was (Gundlach 2005, p. 21):

At the moment, no deviations from ordinary gravity have been found.

James Faller (2005) published an amusing review of measurements of g, the acceleration due to gravity at the surface of the Earth. Faller and his collaborators had previously tested the Fifth Force hypothesis in both Galileo-type falling body experiments and by measuring g as a function of height in a tower. He noted that (Faller 2005, p. 571):

In the end (numerous experiments by many workers later), Newtonian gravity was vindicated.

He also related an amusing anecdote concerning the use of the tower in Erie, Colorado in the tests of the inverse-square law of gravity (p. 571):

NOAA asked a modest $ 1000 in rent for our use of the tower. Their other requirement was that we sign a paper to the effect that if we fell off in the course of making measurements, NOAA would not be held responsible for any personnel free falling due to gravity.

The Eöt-Wash collaboration continued their extensive study of the equivalence principle with a new and improved torsion balance (Schlamminger et al. 2008). Their results for the difference in acceleration for beryllium and titanium test masses, in the northern and western directions, are shown in Fig. 5.19. A violation of the equivalence principle would appear as a difference in the means of the runs taken with the masses in different orientations. The small offset was due to a systematic error, which did not affect their conclusion. Their new upper limits for α, the strength parameter for the Fifth Force or any other deviation from the law

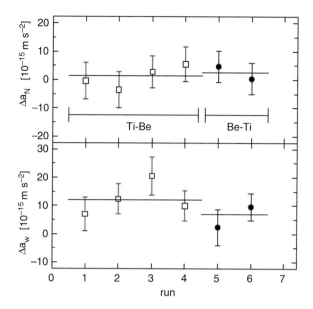

Fig. 5.19 Shown are measured differential accelerations towards north (*top*) and west. After the first four data runs, the Be and Ti test bodies were interchanged on the pendulum frame. A violation of the equivalence principle would appear as a difference in the means (*lines*) of the two data sets. The offset acceleration is due to systematic effects that follow the pendulum frame but not the composition dipole (From Schlamminger et al. 2008)

of gravity are shown in Fig. 5.20. The region of interest for the Fifth Force is at approximately 100 m (p. 041101–1)[21]:

> We used a continuously rotating torsion balance instrument to measure the acceleration difference of beryllium and titanium test bodies towards sources at a variety of distances. Our result $\Delta a_{N,Be-Ti} = (0.6 \pm 3.1) \times 10^{-15}$ m/s^2 improves limits on equivalence-principle violations with ranges from 1 m to ∞ by an order of magnitude. The Eötvös parameter is $\eta_{Earth, Be-Ti} = (0.3 \pm 1.8) \times 10^{-13}$.

Recall that their previous best limit for η was 1.1×10^{-12}. The Fifth Force, if it existed was becoming weaker.

In 2009 two review papers on torsion balance experiments by members of the Eöt-Wash group appeared (Adelberger et al. 2009; Gundlach et al. 2009). The more extensive and detailed report (Adelberger et al. 2009) discussed details and experimental issues involved in torsion balance experiments as well as past experiments and proposed future experiments. The 'Fifth Force' era received only a

---

[21]This was the approximate range suggested in the initial paper, based on the (later withdrawn) results of Stacey and his collaborators. The data of the Eötvös and his collaborators is consistent with ranges up to 1 AU.

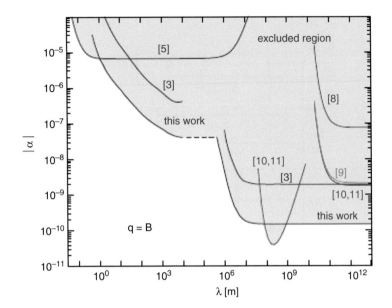

Fig. 5.20 New upper limits on Yukawa interactions coupled to baryon number with 95 % confidence (From Schlamminger et al. 2008)

very brief summary (Adelberger et al. 2009, pp. 108–109):

> After the completion of the classic experiments,[22] little further activity took place until 1986 when Fischbach et al. (1986) reanalysed the Eötvös data. They used this, along with previous claims of anomalous data on g in mines, to claim evidence for a new force. This 'fifth force' was an EP-violating acceleration coupled to B with a range of a few hundred meters that would have rendered it invisible to the classic solar EP tests. This finding triggered many experiments looking for intermediate-range (10 m $< \lambda <$ 10000 km) forces. The Eöt-Wash group at the University of Washington responded by developing a torsion balance mounted on a uniformly rotating platform. The first result from this instrument, which appeared in 1987, ruled out the original fifth force proposal.[23] However, the suggestion of a finite-ranged Yukawa interaction led physicists to broaden their view of EP tests to a search for Yukawa interactions at all accessible length scales.

The second review (Gundlach et al. 2009) restricted itself to experiments done at the University of Washington. It included the results that they had reported in 2008, but no new results.

After 2010 there was very little experimental activity that explicitly dealt with the Fifth Force. This is not to say that there was no work on the related topic of the universality of free fall and tests of the weak equivalence principle. Various experiments conducted in space tested that principle at distances larger than the

---

[22]These were the experiments which test the weak equivalence principle in the fall of bodies toward the Sun, viz., Braginskii and Panov (1972) and Roll et al. (1964).

[23]As we saw in Part I and in the history presented above, this is not accurate.

range of the Fifth Force and there were laboratory experiments that investigated the law of gravity at much smaller distances. An entire issue of *Classical and Quantum Gravity* (Volume 29, Issue 18, 2012), was devoted solely to tests of the weak equivalence principle. One of the articles proposed using atomic interferometry in a free fall from the top of the 141 m high Bremen tower (Hermann et al. 2012, p. 10):

> Ultimately, a drop tower test of the WEP may thus be capable of demonstrating a differential single shot resolution of the order of $\delta a = 10^{-10}g$. Based on such a single shot performance we extrapolate that from about 30 data points a sensitivity estimate on the Eötvös parameter of the order of $\eta < 5 \times 10^{-11}$ may ultimately be obtained. While this is not competitive with current torsion pendulum experiments, and a careful study of systematics would certainly require substantially more than 30 drops, it could already exceed the sensitivity of the current best Galilean free fall test by Kuroda et al.

The Eöt-Wash group paper in that volume did report a new result (Wagner et al. 2012). In addition to their previous result of $\Delta a_{N,Be-Ti} = (0.6 \pm 3.1) \times 10^{-15}$ m/s^2, they presented a new result for an aluminum–beryllium pair, $\Delta a_{N,Be-Al} = (-1.2 \pm 2.2) \times 10^{-15}$ m/s^2. Clifford Will (2014) summarized the situation with respect to the Fifth Force in an extensive review entitled: "The Confrontation between General Relativity and Experiment." He concluded that (Will 2014, p. 27):

> A consensus emerged that there was no credible evidence for a fifth force of nature, of a type and range proposed by Fischbach et al.

Will's summary is, as we have seen, accurate.

5.3 Discussion

There is very strong and persuasive evidence that the fifth Force, as initially proposed by Ephraim Fischbach and his collaborators, does not exist. Numerous experiments have not shown the presence of any force with strength approximately one percent that of Newtonian gravity and with a range of about 100 m (see comment in Footnote 21). Nevertheless I believe the hypothesis has been quite fruitful. It encouraged renewed interest in tests of general relativity, particularly on the weak equivalence principle and on Newtonian gravity at both very large and very small distances and on its composition dependence. As the Eöt-Wash group remarked (Adelberger et al. 2009, pp. 108–109):

> After the completion of the classic experiments,[24] little further activity took place until 1986 when Fischbach et al. (1986) reanalysed the Eötvös data. They used this, along with previous claims of anomalous data on g in mines, to claim evidence for a new force. This 'fifth force' was an EP-violating acceleration coupled to B with a range of a few hundred meters that would have rendered it invisible to the classic solar EP tests. This finding triggered many experiments looking for intermediate-range (10 m $< \lambda <$ 10 000 km) forces. The Eöt-Wash group at the University of Washington responded by developing a

---

[24] See Footnote 22.

torsion balance mounted on a uniformly rotating platform. [...] The first result from this instrument, which appeared in 1987, ruled out the original fifth force proposal.[25] However, the suggestion of a finite-ranged Yukawa interaction led physicists to broaden their view of EP tests to a search for Yukawa interactions at all accessible length scales.

This work also led to improvements in both experimental apparatuses and experimental analyses. As Gillies remarked in 1997 (Gillies 1997, p. 200):

> The contemporaneous suggestion by Fischbach et al. (1986) that there may be previously undiscovered, weak, long-range forces in nature provided further impetus for investigating the composition- and distance-dependence of gravity, since the presence of any such effect might reveal the existence of a new force. [...] Even though convincing evidence in favour of such new weak forces was never found, the many resulting experiments, when viewed as tests of the universality of free-fall, did much to improve the experimental underpinnings of the weak equivalence principle (WEP) of general relativity. In fact, searches for departures from the inverse square behaviour of Newtonian gravity have now come to be interpreted as attempts to uncover violations of the WEP.

As discussed in Part I, some scholars have suggested that the Fifth Force hypothesis should never have been further investigated. These after-the-fact judgments are, I believe, incorrect. As mentioned above the hypothesis was quite fruitful. In addition, I believe that it is important to recognize that wrong science is not bad science. The fact that the Fifth Force hypothesis turned out to be incorrect is not a good reason for saying that it should not have been further investigated. There was, at the time, plausible evidence from the reanalysis of the Eötvös experiment, from the discrepancy between laboratory and mineshaft measurements of g, and from the tantalizing energy dependence of the K^0 decay parameters, that was consistent with the hypothesis. Although one might argue that it was an unlikely hypothesis, the history of science has shown that on occasion such hypotheses have turned out to be correct. Consider the case of parity nonconservation. Distinguished scientists such as Wolfgang Pauli and Richard Feynman were willing to bet that the suggestion by Lee and Yang that parity was not conserved in the weak interactions was incorrect. Feynman bet Norman Ramsey 50 to 1 that parity would be conserved. When experiments showed that parity was not conserved, Feynman paid (for details see Franklin 1986, Chap. 1).

The episode of the Fifth Force is an illustration of good science. A speculative hypothesis, one with some evidential support, was proposed. Further experimentation demonstrated that the hypothesis was incorrect. It did, however, lead to further experimental and theoretical work and improvements in experiments.

---

[25] As we saw in Part I and in the history presented above, this is not accurate.

References

Achilli, V., Baldi, P., et al.: A geophysical experiment on Newton's inverse-square law. Il Nuovo Cim. **112**(B), 775–803 (1997)

Adelberger, E., Gundlach, J.H., et al.: Torsion balance experiments: a low-energy frontier of particle physics. Prog. Part. Nucl. Phys. **62**, 102–134 (2009)

Adelberger, E.G., Heckel, B.R., et al.: Searches for new macroscopic forces. Annu. Rev. Nucl. Part. Sci. **41**, 269–320 (1991)

Adelberger, E.G., Stubbs, C.W., et al.: Testing the equivalence principle in the field of the Earth: particle physics at masses below 1 μeV. Phys. Rev. D **42**, 3267–3292 (1990)

Baldi, P., Campari, E.G., et al.: Testing Newton's inverse square law at intermediate scales. Phys. Rev. D **64**, 082001-1–082001-7 (2001)

Bennett, W.R.: Modulated-source Eötvös experiment at little goose lock. Phys. Rev. Lett. **62**, 365–368 (1989)

Bennett, W.R.: Hunting the fifth force on the Snake river. In: Budker, D., Bucksbaum, P.H., Freedman, S.J. (eds.) Art and Symmetry in Experimental Physics. American Institute of Physics Conference Proceedings Series, vol. 596, pp. 123–155. American Institute of Physics, Melville, NY (2001)

Braginskii, V.B., Panov, V.I.: Verification of the equivalence of inertial and gravitational mass. JETP Lett. **34**, 463–466 (1972)

Carusotto, S., Cavasinni, V., et al.: Test of g universality with a Galileo type experiment. Phys. Rev. Lett. **69**, 1722–1725 (1992)

Carusotto, S., Cavasinni, V., et al.: Limits on the violation of G-universality with a Galileo-type experiment. Phys. Lett. A **183**, 355–358 (1993)

Carusotto, S., Cavasinni, V., et al.: g-universality test with a Galileo-type experiment. Il Nuovo Cim. **111**(B), 1259–1275 (1996)

Cornaz, A., Hubler, B., et al.: Determination of the gravitational constant at an effective interaction distance of 112 m. Phys. Rev. Lett. **72**, 1152–1155 (1994)

Cruz, J.Y., Harrison, J.C., et al.: A test of Newton's inverse square law of gravitation using the 300-m tower at Erie, Colorado. J. Geophys. Res. **96**, 20073–20092 (1991)

Darwin, G.H.: Periodic orbits. Acta Math. **21**, 99–242 (1897)

Dermott, S.F., Murray, C.D.: The dynamics of tadpole and horseshoe orbits: I. Theory. Icarus **48**, 1–11 (1981)

Dittus, H., Mehls, C.: A new experimental baseline for testing the weak equivalence principle at the Bremen drop tower. Class. Quantum Gravity **18**, 2417–2425 (2001)

Faller, J.E.: The measurement of little g: a fertile ground for precision measurement science. J. Res. Nat. Inst. Stand. Technol. **110**, 559–581 (2005)

Fischbach, E., Aronson, S.H., et al.: Reanalysis of the Eotvos experiment. Phys. Rev. Lett. **56**, 3–6 (1986)

Fischbach, E., Talmadge, C.: Six years of the fifth force. Nature **356**, 207–215 (1992)

Fitch, V.L., Isaila, M.V., et al.: Limits on the existence of a material-dependent intermediate-range force. Phys. Rev. Lett. **60**, 1801–1804 (1988)

Focardi, S.: Newton's gravitational law. In: Cianci, R., Collina, R., Francaviglia, M., Fre, P. (eds.) Recent Developments in General Relativity, Genoa 2000, pp. 417–421. Springer, Genoa (2002)

Franklin, A.: The Neglect of Experiment. Cambridge University Press, Cambridge (1986)

Franklin, A.: Gravity waves and neutrinos: the later work of Joseph Weber. Perspect. Sci. **18**, 119–151 (2010)

Gillies, G.T.: The Newtonian gravitational constant: recent measurements and related studies. Rep. Prog. Phys. **60**, 151–225 (1997)

Gundlach, J.H.: Laboratory test of gravity. New J. Phys. **7**, 205-1–205-22 (2005)

Gundlach, J.H., Schlamminger, S., et al.: Laboratory tests of the equivalence principle at the university of Washington. Space Sci. Rev. **148**, 201–216 (2009)

Gundlach, J.H., Smith, G.L., et al.: Short-range test of the equivalence principle. Phys. Rev. Lett. **78**, 2523–2526 (1997)

Heckel, B., Adelberger, E., et al.: Results on the strong equivalence principle, dark matter, and new forces. Adv. Space Res. **25**, 1225–1230 (2000a)

Heckel, B.R., Adelberger, E.G., et al.: Torsion balance test of coupled forces. In: Kursunoglu, B.N., Mintz, S.L., Perlmutter, A. (eds.) Quantum Gravity, Generalized Theory of Gravitation, and Superstring Theory-Based Unification, pp. 153–160. Plenum, New York (2000b)

Hermann, S., Dittus, H., et al.: Testing the equivalence principle with atomic interferometry. Class. Quantum Gravity **29** (18), 184003-1–184003-12 (2012)

Kuroda, K., Mio, N.: Limits on a possible composition-dependent force by a Galilean experiment. Phys. Rev. D **42**, 3903–3907 (1990)

Liu, Y.-C., Yang, X.-S., et al.: Testing non-Newtonian gravitation on a 320 m tower. Phys. Lett. A **169**, 131–133 (1992)

Moffat, J.W., Gillies, G.T.: Satellite Eötvös test of the weak equivalence principle for zero-point vacuum energy. New J. Phys. **4**, 92.1–92.6 (2002)

Moore, M.W., Boudreaux, A., et al.: Testing the inverse-square law of gravity: a new class of torsion pendulum null experiments. Class. Quantum Gravity **6**, A97–A117 (1994)

Moore, G.I., et al.: Determination of the gravitational constant at an effective mass separation of 22 m. Phys. Rev. D **38**, 1023–1029 (1988)

Niebauer, T.M., McHugh, M.P., et al.: Galilean test for the fifth force. Phys. Rev. Lett. **59**, 609–612 (1987)

Nobili, A.M., Bramanti, D., et al.: The 'Galileo–Galilei' (GG) project: testing the equivalence principle in space and on Earth. Adv. Space Res. **25**, 1231–1235 (2000)

Pace, E., De Martini, F., et al.: A capacitive detector to test the principle of equivalence in a free fall experiment. Rev. Sci. Inst. **63**, 3112–3119 (1992)

Reasenberg, R.D., Phillips, J.D.: Testing the equivalence principle on a trampoline. Class. Quantum Gravity **18**, 2435–2445 (2001)

Roll, P.G., Krotkov, R., Dicke, R.: The equivalence of inertial and passive gravitational mass. Ann. Phys. **26**, 442–517 (1964)

Romaides, A.J., Sands, R.W., et al.: Second tower experiment: further evidence for Newtonian gravity. Phys. Rev. D **50**, 3608–3613 (1994)

Romaides, A.J., Sands, R.W., et al.: Final results from the WABG tower gravity experiment. Phys. Rev. D **55**, 4532–4536 (1997)

Sanders, A.J., Deeds, W.E.: Proposed new determination of the gravitational constant G and tests of Newtonian gravitation. Phys. Rev. D **46**, 480–504 (1992)

Schlamminger, S., Choi, K.Y., et al.: Test of the equivalence principle using a rotating torsion balance. Phys. Rev. Lett. **100**, 041101-1–041101-4 (2008)

Slobodrian, R.J.: Study of new fundamental forces in a microgravity environment. Class. Quantum Gravity **9**, 1115–1119 (1992)

Smith, G.L., Hoyle, C.D., et al.: Short-range tests of the equivalence principle. Phys. Rev. D **61**, 022001-1–022001-20 (2000a)

Su, Y., Heckel, B.R., et al.: New tests of the universality of free fall. Phys. Rev. D **50**, 3614–3636 (1994)

Thieberger, P.: Search for a substance-dependent force with a new differential accelerometer. Phys. Rev. Lett. **58**, 1066–1069 (1987c)

Unnikrishnan, C.S.: Search for a 5th force. Pramana J. Phys. **41**(Supplement S), 395–411 (1993)

Venema, B.J., Majumder, P.K., et al.: Search for a coupling of the Earth's gravitational field to nuclear spins in atomic mercury. Phys. Rev. Lett. **68**, 135–138 (1992)

Wagner, T.A., Schlamminger, S., et al.: Torsion-balance tests of the weak equivalence principle. Class. Quantum Gravity **29**(18), 184002-1–184002-15 (2012)

Will, C.W.: The confrontation between general relativity and experiment. Living Rev. Relativ. **17**, 1–117 (2014)

Yang, X., Liu, W., et al.: Testing the intermediate-range force at separations around 50 meters. Chin. Phys. Lett. **8**, 329–332 (1991)

Chapter 6
The Fifth Force: A Personal History, by Ephraim Fischbach

6.1 Introduction

At approximately 11 AM on Monday, January 6, 1986 I received a call from John Noble Wilford of the *New York Times* inquiring about a paper of mine which had just been published in *Physical Review Letters* (PRL). As a subscriber to the *Times* I knew who John was, and so it was exciting to find myself speaking to him in person. My excitement was tempered by the fact that I had returned the day before to Seattle with a major cold which made it difficult for me to talk to him or anybody else. Two days later a front page story appeared in the *Times* by John under the headline "Hints of 5th Force in Universe Challenge Galileo's Findings," accompanied by a sketch of Galileo's supposed experiment on the Leaning Tower of Pisa. Thus was born the concept of a "fifth force". As used now, this generically refers to a gravity-like long-range force (i.e., one whose effects extend over macroscopic distances) co-existing with gravity, presumably arising from the exchange of any of the ultra-light quanta whose existence is predicted by various unification theories such as supersymmetry. Depending on the specific characteristics of this hypothesized force, it could manifest itself in various experiments as an apparent deviation from the predictions of Newtonian gravity.

Our paper in *Physical Review Letters* entitled "Reanalysis of the Eötvös Experiment" (Fischbach et al. 1986a), was co-authored by my three graduate students Carrick Talmadge, Daniel Sudarsky, and Aaron Szafer, along with my long-time friend and collaborator Sam Aronson. As the title suggests, our paper re-analyzed the data obtained from what is now known as the "Eötvös Experiment", one of the most well-known experiments in the field of gravity (Eötvös et al. 1922; Szabó 1998). The authors of that 1922 paper, Baron Loránd Eötvös, Desiderius

Reprinted with permission from "The fifth force: A personal history, Ephraim Fischbach, Eur. Phys. J. H, (2015), © EDP Sciences, Springer-Verlag 2015.

© Springer International Publishing Switzerland 2016
A. Franklin, E. Fischbach, *The Rise and Fall of the Fifth Force*,
DOI 10.1007/978-3-319-28412-5_6

Pekár, and Eugen Fekete (EPF), had carried out what was then the most precise test of whether the behavior of objects in a gravitational field was the same independent of their different chemical compositions. Their conclusion, that it was the same to approximately one part in 10^9, provided experimental support for what is now known as the Weak Equivalence Principle (WEP), which is one of the key assumptions underlying Einstein's General Theory of Relativity (Will 1993). However, the result of our reanalysis of the EPF paper (Eötvös et al. 1922; Szabó 1998) was that the EPF data were in fact "sensitive to the composition of the materials used," in contrast to what EPF themselves had claimed. If the EPF data and our reanalysis of them were both correct, then one implication of our paper would be that EPF had discovered a new "fifth force" in nature.

Approximately 30 years have elapsed since the publication of our PRL, and we now know with a great deal of confidence that a "fifth force" with the attributes we assumed does not exist. We can also exclude a large number of generalizations of the original fifth force hypothesis by noting that, at present, there is no evidence for any new force beyond the established strong, electromagnetic, weak (or electroweak) and gravitational forces. Among the many things we do not know is what EPF could have done in their classic experiment to have delivered to us (some six decades later) evidence at the ~ 8 standard deviation (8σ) level for a new force with attributes that could not have even been conceptualized at that time.

As discussed in the epilogue of Sect. 6.8, it is, of course, possible that EPF did everything correctly, in which case our apparent failure to understand, and thereby reproduce, their results may be our fault not theirs. The fifth force story is thus a continuing one, in which its past will certainly inform its future. This story is also of interest in that it provides yet another example of how the scientific community gives birth to an idea, tests it, and then accepts or rejects it based on the results of experiment.

My objective here is to present the fifth force story as I experienced it personally, from its inception to the present. My task has been greatly simplified by the existence of Allan Franklin's history, *The Rise and Fall of the Fifth Force* (Franklin 1993), which gives a detailed annotated history of the fifth force effort along with extensive references. Several other sources will also be helpful. In 1999 Carrick Talmadge and I published a detailed technical description of fifth force searches under the title *The Search for Non-Newtonian Gravity* (Fischbach and Talmadge 1999). In preparation for this book we felt it appropriate to compile a formal bibliography of more than 800 experimental and theoretical papers related to the fifth force searches prior to 1992 which was published in the journal *Metrologia* (Fischbach et al. 1992). Since the central focus of this review will be on our reanalysis of the EPF paper, I will also make reference to the much expanded version of our original paper which appeared in 1988 in *Annals of Physics* (Fischbach et al. 1988), which is briefly outlined in Appendix 1.

In order to streamline the fifth force narrative, I have provided additional technical background in the appendices when needed. As noted above, Appendix 1 contains a brief summary of the fifth force formalism, and Appendix 2 presents the phenomenology of the K^0–$\overline{K}^0$ system. Appendices 3, 4, and 5 present, respectively, historically interesting correspondence from Robert Dicke, *Physical Review Letters*, and Richard Feynman. Appendix 6 relates to one of the lighter moments in the fifth force saga.

Let me conclude by apologizing in advance to my many friends and colleagues whose contributions, for reasons of space, I have not been able to discuss here. The history covered here focuses on small parts of the story which were significant to me personally at the time for various reasons. It is my hope that in the references cited here, especially in Allan's book (Franklin 1993), our book (Fischbach and Talmadge 1999) and the accompanying Metrologia bibliography (Fischbach et al. 1992), they will receive the full credit they genuinely deserve.

6.1.1 Brief History

In tracing back the body of work now known by the generic rubric "fifth force", it is natural for historians to ask "where and how did it all begin?" The answer to "where" is relatively straightforward: it began at my home institution Purdue, motivated in large measure by the beautiful, Colella, Werner, Overhauser (COW) experiment in 1975 (Colella et al. 1975) to be discussed below, followed by sabbaticals at the Institute for Theoretical Physics (ITP, now C.N. Yang ITP) at Stony Brook (1978–1979), and at the Institute for Nuclear Theory at the University of Washington (1985–1986).

The "how" is less obvious, and consequently much more interesting. In broad outlines, to be fleshed out below, the COW experiment which tested the validity of Newtonian gravity at the quantum level, led me to pursue the question of whether we could test Einstein's theory of General Relativity (GR) at the quantum level. In considering the possibility of alternatives to GR at the quantum level, I was implicitly considering the possibility that new forces existed in nature whose presence had not yet been detected. This was the focus of much of my work at ITP-Stony Brook during my (1978–1979) sabbatical, and led to several publications (Fischbach 1980; Fischbach and Freeman 1980; Fischbach et al. 1981), including an award for an essay submitted to the Gravity Research Foundation (Fischbach and Freeman 1979).

However, my research at Stony Brook produced a surprise as a result of a collaboration with Sam Aronson related to an anomalous energy dependence he was detecting in Fermilab data on neutral kaons. When produced in strong interactions, the neutral kaon K^0 and its antiparticle $\overline{K}^0$ are distinguished by the strangeness quantum number, $S = +1$ and $S = -1$, respectively. However, when they decay via the weak interaction strangeness is not conserved, and this results in a mixing of K^0 and $\overline{K}^0$ to form two new neutral states K_L^0 and K_S^0. These are eigenstates of the full Hamiltonian, and their decays follow the usual exponential decay law with K_L^0 (K_S^0) being the longer- (shorter-) lived state. The K_L^0–K_S^0 system is thus described by the mean lifetimes τ_L and τ_S of the two states, and their (slightly different) masses m_L and m_S. Additionally, the observation of CP-violation in the K_L^0–K_S^0 system introduces the parameters η_{+-} and η_{00} which characterize, respectively, the amplitudes for the CP-violating decays $K_L^0 \to \pi^+\pi^-$ and $K_L^0 \to \pi^0\pi^0$. As explained below, these data hinted at the possible presence of a new force, and hence my research during the period 1979–1985 focused heavily on analyzing these data, as well as on my ongoing interest in tests of GR at the quantum level.[1]

In August 1985 I traveled with my family to the University of Washington (UW) in Seattle to spend a year-long sabbatical at the Institute for Nuclear Theory in the Department of Physics. I was accompanied by one of my three graduate students, Carrick Talmadge, for whom our eventual reanalysis of the Eötvös experiment would become the subject of his Ph.D. dissertation. I had been working up to that point with Norio Nakagawa at Purdue on possible modifications of the electron anomalous magnetic moment $(g - 2)$ arising from the suppression of some electromagnetic vacuum fluctuations due to the $(g - 2)$ apparatus (Fischbach and Nakagawa 1984a,b). (This is vaguely similar to the well-known Casimir effect.) We had submitted our latest paper for publication, but the reviewer wanted us to carry out some additional calculations, which neither of us was interested in doing. So I turned my attention instead to studying neutral kaon experiments as probes for new long-range forces.

There was no compelling evidence then (nor is there any now) for new long-range forces. Hence the best that kaon experiments (or any other experiment) can do is to constrain the magnitudes of the various parameters that would characterize such a force in a particular theory. As we discuss below, a very useful compilation of such constraints was published in 1981 by Gibbons and Whiting (GW) (1981), based on an elegant formalism developed by Fujii (1971, 1972, 1974). However, the implications of the classic 1922 paper by Eötvös, Pekár, and Fekete (EPF) were not included, and neither were the similar experiments of Roll, Krotkov, and Dicke (RKD) (1964), or Braginskii and Panov (BP) (1972), for reasons to be discussed below. The ABCF series of papers (Aronson et al. 1982, 1983a,b; Fischbach et al. 1982) written by Sam Aronson, Greg Bock, Hai-Yang Cheng, and me had yet to appear at the time of the GW paper, and hence there was additional information on possible long-range forces yet to be incorporated into an overall set of constraints on new forces. As will become clear shortly, these constraints taken together would become central in our analysis of the EPF experiment.

My sabbatical at the University of Washington had been arranged by Wick Haxton whom I knew from the time when he was an Assistant Professor at Purdue. Wick was also the colleague who brought to my attention the work of Frank Stacey and Gary Tuck (Stacey 1978, 1983; Stacey et al. 1981; Stacey and Tuck 1981, 1984) in Australia. Frank and Gary had determined the Newtonian gravitational constant G as measured in a deep mine and found that it was larger than the standard laboratory value G_0 by approximately 0.5 %–1.5 %. One possible explanation of this

---

[1] For further discussion of the K_L^0–K_S^0 system, see Appendix 2.

difference would be a new long-range force whose influence would extend over a limited distance scale of a few kilometers. As noted in our paper (Fischbach et al. 1986a) (see also Appendix 1), such a force could be described by introducing a non-Newtonian interaction of the form

$$V(r) = -G_\infty \frac{m_1 m_2}{r}\left(1 + \alpha e^{-r/\lambda}\right) \equiv V_N(r) + \Delta V(r) \,, \tag{6.1}$$

where $V_N(r)$ is the usual Newtonian potential energy for two masses m_1 and m_2 separated by a distance r. In a private communication from Frank Stacey he noted that the discrepancy that he and Tuck had found could then be explained if α and λ had the values

$$\alpha = -(7.2 \pm 3.6) \times 10^{-3} \,, \qquad \lambda = 200 \pm 50 \,\mathrm{m} \,. \tag{6.2}$$

Upon examining the paper by GW (Gibbons and Whiting 1981) in more detail, I recognized that an interaction characterized by (6.1) and (6.2) with the indicated values of α and λ was in fact reasonably compatible with then-existing data. Moreover, the RKD and BP results, which did not appear in the GW paper, were also compatible with (6.1) and (6.2), and hence the only remaining experiment which could rule out a new force characterized by (6.1) and (6.2) was the original EPF experiment. This realization then became the proximate motivation for our reanalysis of the EPF experiment, and our discovery in the EPF paper of evidence for what shortly became known as the "fifth force".

From the preceding discussion it may seem at first surprising that the earlier (and less sensitive) EPF experiment became the focus of my attention, rather than the similar (but much more sensitive) RKD and BP experiments. The reason for this is that the later experiments achieved their increased sensitivity in part by measuring the acceleration differences of two samples to the Sun, whereas EPF compared the accelerations of their samples under the influence of the Earth's gravitational field. Using the Sun as a source allowed the daily rotation of the Earth to modulate any potential signal in a way that suppressed possible systematic errors. In contrast, EPF resorted to physically rotating their apparatus in the laboratory to suppress effects such as intrinsic twists in their torsion fibre. However, this also had the unwanted effect of disturbing the fibre itself, which RKD and BP sought to avoid.

Since the Sun was the presumed source of any possible acceleration difference of the test masses used in either the RKD or BP experiments, a force emanating from the Sun whose range λ was only of order 200 m, would have no influence on any terrestrial experiment. This follows from (6.1) by noting that $e^{-r/\lambda}$ is immeasurably small when $r = 1.5 \times 10^8$ km is the Earth–Sun distance and $\lambda \approx 200$ m. Hence, the EPF experiment remained as the only potential obstacle to formulating a theory based on (6.1) and (6.2) which could potentially account for both the anomaly detected by Stacey and Tuck, and the anomalous energy dependence the kaon regeneration data that Sam, Greg, Hai-Yang, and I had published.

However, one last question remained before I was willing to commit myself and Carrick Talmadge to the time-consuming effort of re-examining the EPF experiment in detail. That was making absolutely certain that the presumed source of any effect in the EPF experiment was in fact the Earth and not the Sun. I was much more familiar with both the RKD and BP experiments because I had used their data just a year earlier in a paper co-authored with Hai-Yang Cheng, along with Mark Haugan and Dubravko Tadić (Fischbach et al. 1985). This paper, which established an interesting connection between Lorentz-Noninvariance and the Eötvös experiments, did not actually use the EPF data, but only the more sensitive RKD and BP results.

Because I do not read German I enlisted the help of Peter Buck who was a postdoc at INT from Germany. I tasked him initially with answering the question of whether EPF were comparing the accelerations of objects falling to the Earth, which he did in the affirmative. Eventually Peter's effort extended to a full-translation of the EPF paper as we describe below.

Having convinced myself that the EPF experiment was the only remaining impediment to postulating the existence of a new force capable of explaining both the anomalous energy dependence of the neutral kaon parameters, and the anomalies found by Stacey and Tuck, I set about the task of re-analyzing the EPF paper. Not surprisingly, the trajectory that began in 1975 with my focus on the COW experiment and quantum gravity, and which ultimately led through kaon physics to the EPF experiment, was more complicated than suggested by this brief outline. The remainder of this Introduction will thus be devoted to filling in these missing details, some of which were crucial in leading to our reanalysis of the EPF experiment and the fifth force hypothesis.

6.1.2   The COW Experiment and Its Impact

As noted above, in 1975 my colleagues Roberto Colella and Al Overhauser published a remarkable paper which provided much of the original motivation for my subsequent work leading to our group's reanalysis of the EPF experiment. In this paper the authors showed that one could carry out an experiment which tested the quantum behavior of neutrons in a gravitational field. Not long thereafter they were joined by Sam Werner in actually carrying out this experiment (Colella et al. 1975), now known as the COW experiment, in which they verified experimentally that the quantum-mechanical behavior of nonrelativistic neutrons in a weak gravitational field agreed with theoretical expectations based on Newtonian gravity and the Schrödinger equation. (The original apparatus is now on display in the Physics and Astronomy library at Purdue.)

This pioneering experiment had only one shortcoming from my point of view, and it is best illustrated by an anecdote that Al told relating to the time he gave a lecture on this experiment at Brookhaven National Laboratory. When he got to the conclusion that the COW results were in agreement with predictions (assuming Newtonian gravity and the Schrödinger equation), Maurice Goldhaber commented to the effect that "... of course they do, if they didn't we would never have allowed you to publish them!" The content of Goldhaber's comment was clear: since both Newtonian gravity and the Schrödinger equation have been so well tested, and that is all that is needed to derive the theoretical prediction for the COW effect, there is no way COW could have obtained any other result. Thus, although the COW experiment is a genuine test of gravity at the quantum level, it did not test gravity in a way that would provide much insight into how to formulate a truly quantum theory of gravity, a problem which remains unsolved to this day.

Al's office was just a few doors down from my own, and we talked very often about subjects of mutual interest, especially about the COW experiment and its implications. Al was convinced that the observed CP-violation in the K^0–$\overline{K}^0$ system was due to some external gravity-like field, and in one conversation we had early in the "COW era" he made a comment which eventually led me to the following observation. In the Earth's gravitational field, consider the energy difference between a K_L^0 and K_S^0 (whose mass difference is $\Delta m = m_L - m_S$) over a vertical height $\hbar/c\Delta m$. This energy difference is given by $m_K g(\hbar/c\Delta m)$, where $m_K = (m_L + m_S)/2$, and $g = 980 \,\text{cm/s}^2$. (This vertical distance is that which a virtual relativistic kaon would travel in a time $t = \hbar/c^2\Delta m$.) If we compare this energy difference to the mass-energy difference of K_L and K_S, we find (Fischbach 1980)

$$\frac{m_K g(\hbar/c\Delta m)}{c^2 \Delta m} \approx 0.84 \times 10^{-3} \, . \tag{6.3}$$

This is tantalizingly close to the magnitude of the CP violating parameter $\mathrm{Re}\, \varepsilon/2 = (0.80 \pm 0.01) \times 10^{-3}$ (PDG 2014). Although this may be no more than a surprising coincidence, it certainly provided part of our subsequent motivation to somehow connect anomalies in the K^0–$\overline{K}^0$ system with gravity via the EPF experiment.

Since kaon experiments are inherently relativistic, the suggestion of (6.3) that there could be a connection between gravity and CP-violation in the K^0–$\overline{K}^0$ system led me to ask whether we could design a relativistic analog of the COW experiment. In contrast to the COW experiment itself, which only tested Newtonian gravity, such a relativistic experiment could in principle test some aspects of Einstein's General Theory of Relativity (GR) and various alternatives to GR. Stated another way, a relativistic experiment could test whether the parametrized post-Newtonian (PPN) parameters $\alpha_{\mathrm{PPN}}, \beta_{\mathrm{PPN}}, \gamma_{\mathrm{PPN}}, \ldots$, which characterized the metric tensor in the weak-field limit at the macroscopic level, were the same as would describe the metric tensor at the quantum level. At the macroscopic level these parameters are defined in the terms of the components of metric tensor $g_{\mu\nu}(x)$ for a spherically symmetric geometry expressed in isotropic coordinates. To lowest order in $\Phi = GM_\odot/cr^2$,

$$ds^2 = f(r)\left(dx^2 + dy^2 + dz^2\right) + g_{00}(r)\left(dx^0\right)^2 \, , \tag{6.4}$$

where

$$r = \left(x^2 + y^2 + z^2\right)^{1/2} . \tag{6.5}$$

The metric components $f(r)$ and $g_{00}(r)$ are then given by

$$f(r) = 1 + 2\gamma_{PPN}\Phi + \frac{3}{2}\delta_{PPN}\Phi^2 + \mathcal{O}(\Phi^3) , \tag{6.6}$$

$$-g_{00}(r) = 1 - 2\Phi + 2\beta_{PPN}\Phi^2 + \mathcal{O}(\Phi^3) . \tag{6.7}$$

The utility of the PPN formalism is that it allows the predictions of various theories of gravity to be readily inter-compared in terms of a common set of PPN parameters (Will 1993). Going further, we can reproduce some classic predictions of GR at the macroscopic level without even knowing much about GR at all (Fischbach and Freeman 1980). For example, the gravitational deflection of light by the Sun can be calculated as a classical geometric optics problem by noting that a photon can be viewed as propagating in a Minkowskian space-time but with a local index of refraction

$$n(r) = [-f(r)/g_{00}(r)]^{1/2} . \tag{6.8}$$

It seemed to me that, absent such basic information, it would be difficult to make rapid progress in formulating a truly quantum theory of gravity. As but one example, this would address to some extent the question of whether gravity at the macroscopic level was merely an effective theory, where the PPN parameters were appropriate averages over some other parameters which would characterize space-time at the quantum level.

From many points of view the K^0–$\overline{K}^0$ system would be an ideal choice to pursue this question because relativistic kaons exhibit interference phenomena which are clear indications of quantum behavior (Aronson et al. 1982, 1983a,b; Fischbach et al. 1982). Studying the behavior of kaons in a weak gravitational field would thus be a quantum analog of the deflection of light passing the Sun. This is the famous Eddington experiment which brought world-wide fame to Einstein by demonstrating (in modern terminology) that γ_{PPN} was indeed close to 1 as predicted by GR.

There is, however, a fundamental problem with the K^0–$\overline{K}^0$ system, and it is the very feature which makes it interesting. In order to carry out an analog of the COW experiment one would have to coherently split a kaon beam in a gravitational field and then recombine the split beams after they had traveled along different paths in the field. For the low-energy neutrons which were used in the COW experiment, their de Broglie wavelengths were comparable to the silicon lattice spacing in the crystal used. Hence the lattice could coherently split the neutrons, just as it would an X-ray beam of comparable wavelength. This splitting of the neutron beam with wavelength λ then produces a phase shift $\Delta\phi$ of the two components given by

$$|\Delta\phi| = \frac{2\pi m_n^2 g \ell_1 \ell_2 \lambda}{h^2}, \qquad (6.9)$$

where m_n is the neutron mass, $g = 980\,\mathrm{cm/s^2}$, ℓ_1 is the linear distance they travel, and ℓ_2 is the vertical separation. In the original COW experiment $A = \ell_1 \ell_2 \approx 10\,\mathrm{cm^2}$ was the macroscopic area enclosed by the split beams, and this leads to a macroscopically observable signal. However, the de Broglie wavelength of a relativistic kaon is so small that splitting it via any atomic lattice is not feasible. For example, the de Broglie wavelength of a kaon with momentum $10\,\mathrm{GeV}/c$ is approximately $10^{-6}\,\mathrm{\mathring{A}}$, which is much smaller than any atomic lattice spacing. However, the preceding discussion does not entirely preclude tests of GR at the quantum level, and an example of such an experiment is given in Fischbach (1984). Consider the process

$$e^+ + e^- \longrightarrow \phi(1020) \longrightarrow K_L^0 + K_S^0, \qquad (6.10)$$

where both K_S^0 and K_L^0 can decay into $\pi^+\pi^-$, the latter by virtue of CP-violation. In the absence of gravity various symmetry arguments constrain the form of the $2(\pi^+\pi^-)$ final state. However, in the presence of gravity these final states are perturbed in a manner that could allow for a test of GR at the quantum level. The difficulty with carrying out such an experiment in practice is that for $\phi(1020) \to K_L^0, K_S^0$, the outgoing K_L^0, K_S^0 are nonrelativistic and hence this particular decay mode is not particularly useful for our purposes. By way of contrast, the K_L^0 and K_S^0 produced in the decay of $J/\Psi(1S)$ would be sufficiently relativistic to provide a meaningful GR test in principle. However, although the final K^0–$\overline{K}^0$ state is one of the dominant decay modes of $\phi(1020)$, it is only a minor decay mode of $J/\Psi(1S)$ decay. Thus the small branching ratio for this mode (2×10^{-4}) precludes at present any meaningful test of GR using the K^0–$\overline{K}^0$ (or K_L^0–K_S^0) system.

6.1.3 Stony Brook Sabbatical (1978–1979)

I had been a research associate at ITP-Stony Brook during the years 1967–1969, and I had been invited to return for my sabbatical. The decision to go on sabbatical was not an easy one for my wife Janie and me: our second son Jeremy was born prematurely in April of 1978, and the thought of moving from Indiana to Stony Brook with the very young children was not appealing. Janie and I had even talked about simply canceling our sabbatical plans entirely. But in the end Janie felt that this sabbatical was important to me, although neither of us could foresee at that time what would eventuate. We were accompanied on my sabbatical by my two graduate students, Hai-Yang Cheng and Belvin Freeman.

The previously discussed difficulty of testing GR at the quantum level, by developing an analog of the COW experiment in the K^0–$\overline{K}^0$ system, eventually led me to consider tests in atomic systems, specifically in hydrogen and positronium. Eventually this became the subject of Belvin's Ph.D. thesis. As is well known, in classical Bohr theory the velocity of an electron in the ground state of hydrogen is $\beta = v/c \approx \alpha = e^2/\hbar c \approx 1/137$. This is sufficiently large to motivate consideration of the possibility of testing GR in hydrogenic systems. My problem was that the requisite calculations involved understanding, and dealing with, the Dirac equation in GR with which I was not familiar. Although I had taught GR, relativistic quantum mechanics, and introductory field theory a number of times, I had never discussed the effects of gravity in relativistic quantum systems. Fortunately for me Fred Belinfante of our department, a noted GR expert, decided to teach GR during the Fall of 1976 prior to my sabbatical, and this included studying the Dirac equation in GR.

Much of the 1978–1979 sabbatical at Stony Brook was devoted to exploring with Belvin possible experimental tests of GR in hydrogen and positronium, using the formalism I had learned from Fred Belinfante. We showed in a series of papers (Fischbach and Freeman 1979; Fischbach 1980; Fischbach et al. 1981) that for a hydrogen atom at rest the Earth's gravitational field produced an analog of the electromagnetic Stark effect, in the sense of mixing unperturbed states of opposite parity. The energy scale for these effects is determined by a constant $\eta = g\hbar/c$, where $g = 980\,\mathrm{cm/s^2}$ is the familiar acceleration of gravity at the surface of the Earth. Not surprisingly, $\eta \to 0$ when either $\hbar \to 0$ or $c \to \infty$, which supports our intuition that we are in fact studying a genuine GR effect at the quantum level. Since $\eta = 2.2 \times 10^{-23}\,\mathrm{eV}$ at the surface of the Earth, and would only be $3.5 \times 10^{-12}\,\mathrm{eV}$ at the surface of a typical neutron star, prospects for directly observing GR effects in hydrogen or positronium are bleak at present. Our summary paper (Fischbach et al. 1981), written in collaboration with Wen-Kwei Cheng at the University of Delaware, made it clear how difficult it is likely to be to detect the presence of GR effects in even the most sensitive atomic systems.

Although my intention at the outset of my Stony Brook sabbatical was to devote myself primarily to testing GR in atomic systems, my research took an unexpected turn after a visit from my friend Sam Aronson, who was then in the Physics Department at Brookhaven National Laboratory, and subsequently rose to be its Chairman. Sam eventually became the Director at Brookhaven, and is the 2015 President of the American Physical Society. Sam and I had known each other from our undergraduate days at Columbia when we were both in the same philosophy of science class at Barnard taught by Daniel Greenberger. The purpose of Sam's visit was to enlist my help in a problem he was having understanding the results of an experiment at Fermilab with which he was involved, along with Val Telegi, Bruce Winstein, Greg Bock, and others. This experiment was aimed at studying the process of K_S^0 regeneration in which K_S^0 mesons could be regenerated from a K_L^0 beam by passing that beam through a target such as hydrogen, carbon, or lead. The experimental results were of interest because there was well-developed formalism

(Regge pole theory) which predicted what this energy dependence should be. (See Appendix 2 for a discussion of kaon regeneration.)

Neutral kaon regeneration is an extremely interesting phenomenon in part because it is an elegant example of quantum mechanical interference. This interference arises from the fact that both K^0_L and the regenerated K^0_S can decay into $\pi^+\pi^-$ (and also $\pi^0\pi^0$). The former decay is CP-violating and is hence suppressed, while the latter decay is CP-allowed but is suppressed by virtue of the fact that the regeneration amplitude is itself small. The net effect is that the decay amplitude of a neutral kaon beam into $\pi^+\pi^-$ arises from the interference between two decay processes with amplitudes which can be roughly comparable. This leads to an oscillatory behavior of the detected $\pi^+\pi^-$ amplitude which is described by a function $\cos[\Delta mt + \phi_\rho(E) - \phi_{+-}]$ where (in units where $\hbar = c = 1$) E is the laboratory energy, and ϕ_{+-} is the phase characterizing the CP-violating $K^0_L \to \pi^+\pi^-$ decay. Knowing E and ϕ_{+-} one can then extract the desired strong interaction phase $\phi_\rho(E)$. Sam's problem was that the energy dependence he and his group were finding at Fermilab was far greater than that expected from theory (Fig. 6.1). (See Appendix 2 for more details.)

Sam and I arranged for us to meet with C.N. Yang, and during this meeting Yang agreed that Sam's data were not compatible with any model that he knew. Sam was analyzing the Fermilab data with his student Greg Bock at the University of Wisconsin, and I was accompanied on my sabbatical by my students Hai-Yang Cheng and Belvin Freeman. Since Hai-Yang had essentially finished his Ph.D. research by that time, I suggested that he and I join forces with Sam and Greg to try to understand the apparently anomalous energy dependence of the Fermilab data.

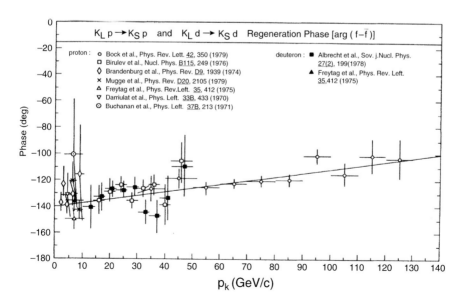

Fig. 6.1 Plot of ϕ_{21} vs. kaon momentum taken from Aronson et al. (1983a)

As it turns out the strong-interaction formalism being used to predict the regeneration phase was Regge pole theory, a subject which I had previously promised myself never to get involved with. Having no choice at this point, I immersed myself in this formalism, and eventually wrote a long appendix to one of our papers (Aronson et al. 1983a) in which we verified that Regge pole theory did in fact predict too small an energy dependence to account for the observed Fermilab data. (This discussion was sufficiently detailed that one of the reviewers of this paper commented that this appendix should have been published as a separate paper.)

Although kaon regeneration would seem to have nothing to do with the COW experiment, gravity, or the eventual search for a fifth force, a pivot point came during a meeting one day among Sam, Greg, Hai-Yang, and me. As noted above, the regeneration phase $\phi_\rho = \phi_\rho(E)$ appeared in the relevant formulas via a factor $\cos[\Delta mt + \phi_\rho(E) - \phi_{+-}]$, where $\Delta m = m_L - m_S$ is the K_L^0-K_S^0 mass difference, and ϕ_{+-} is the phase of the CP-violating parameter η_{+-}. The energy dependence of ϕ_ρ thus depended on assuming (as we all then did) that Δm, η_{+-}, and ϕ_{+-} were fundamental constants of nature, and hence independent of the laboratory energy of the kaon beam that we were studying. (It should be noted that measurements of these parameters are traditionally referred back to the kaon rest frame.) Hence any energy dependence of the combination $(\phi_\rho - \phi_{+-}) \equiv \Phi$ must be due to ϕ_ρ, and this energy dependence was the problem we were facing in light of our Regge pole analysis, along with the work of others.

The pivotal moment came when we started to consider the possibility that ϕ_{+-} itself was energy-dependent, and hence that the energy dependence of Φ was actually due mostly to that of ϕ_{+-}. We recognized that, as unconventional this suggestion was, such an energy dependence could arise from the interaction of the K^0-$\overline{K}^0$ system with some new external field. This was not a new idea, since such an interaction had been proposed independently by Bell and Perring (1964) and independently by Bernstein et al. (1964) to explain CP-violation. However, their formalisms implied that the energy variation of the CP violating parameter $|\eta_{+-}|$ would be quite large (see below), and hence this proposal was quickly ruled out.

Nonetheless, through a study of the energy dependence of $\phi_\rho(E)$, Sam, Greg, Hai-Yang, and I had raised the idea of some sort of new long-range force. This thread would eventually connect to the work of Stacey and Tuck, whose geophysical determination of the Newtonian constant of gravity G found an anomaly, which could also be attributed to the presence of a new force.

Eventually Sam, Greg, Hai-Yang, and I felt sufficiently confident in our analysis that we submitted a paper giving our results to *Physical Review Letters* (PRL). Our original version met with stiff resistance from PRL. Just as it looked as though we would never succeed in publishing these data, not to mention the accompanying theoretical analysis, I had an idea motivated by a Bruegel painting I had studied as an undergraduate at Columbia. In this painting, "Landscape with the Fall of Icarus", Bruegel takes the central purpose of the picture, namely depicting the story of the fall of Icarus escaping from Crete because he flew too close to the Sun, and makes it an incidental detail in an otherwise pastoral scene (Hughes and Bianconi 1967).

So incidental is Icarus' plunge into the sea, that it could easily be missed by someone not familiar with the painting. In fact, on a trip out West many years ago with my family we ended up in a motel room with this painting on the wall. Except that the painting had been cropped to allow it to fit into one of their standard size frames, with the result that Icarus was now completely missing![2]

As applied to our situation at that time, my suggestion to the group was to write a theoretical/phenomenological paper focusing on our formalism in which our actual experimental results appeared to be almost incidental. This stratagem worked, and a phenomenological paper containing our data was accepted relatively quickly by *Physics Letters*, and was published on 30 September 1982 (Fischbach et al. 1982). In the meantime, a rewritten version of our original data and analysis was submitted to PRL and accepted, and was published on 10 May 1982 (Aronson et al. 1982). The acceptance of these papers appeared to break the log jam we were confronting, and full length papers presenting our data and our phenomenological formalism appeared in back-to-back papers in *Physical Review D* (Aronson et al. 1983a,b).

There was, however, a problem remaining in trying to attribute the apparent energy dependence of the K^0–$\overline{K}^0$ parameters to a new external field, namely the experimental evidence that this could not explain CP-violation. A critical turning point came on the evening of December 6, 1983. I had been asked to sit on an NSF panel charged with awarding NATO postdoctoral fellowships, and I was leaving the next morning to San Francisco to join that panel. After dinner I decided to tidy up the notes I was working on during the day as a form of relaxation. Sometime around 10 PM I made what to me was at that time a startling observation in an equation I had just written down. As noted above, it had been shown by Bell and Perring (BP) (Bell and Perring 1964), and simultaneously by Bernstein, Cabibbo, and Lee (BCL) (Bernstein et al. 1964), that if the observed CP violation was due to the interaction of the K^0–$\overline{K}^0$ system with an external source mediated by a quantum ("hyperphoton") that had a spin J (in units of Planck's constant), then the magnitude of the CP-violating parameter η_{+-} should vary with the laboratory energy E (or velocity $\beta = v/c$) of the kaons as γ^{2J}, where $\gamma = E_K/mc^2 = \sqrt{1 - \beta^2}$ is the usual relativistic factor. Since the hyperphoton was presumed to be a vector field ($J = 1$), which was required in such a picture to produce an energy difference between K^0 and its antiparticle $\overline{K}^0$, the expected energy dependence was thus γ^2. Shortly after their proposal experiments searched for a γ-dependence, but found none (De Bouard et al. 1965; Galbraith et al. 1965; Lee and Wu 1966). This was a compelling argument at the time against the hyperphoton mechanism as an explanation of the observed CP-violation. However, what I had observed in the equation I had just written was a cancellation among terms which, for the system I was analyzing, eliminated the term proportional to γ^{2J} leaving a residual term with a much smaller energy dependence. If my algebra was correct, the hypercharge mechanism as an explanation of CP-violation was now again viable.

---

[2]For a literary reference, see W.H. Auden "Musée des Beaux Arts".

The implications of this result were immediately obvious to me, so much so that I could not even write down the next equation, in which the canceling terms would have no longer been present. As a teenager I had played a lot of chess, and so I pictured what had just happened as if I had "checkmated" the problems associated with the hypercharge mechanism. I went to sleep and arranged to awaken at 4 AM the next morning to check my algebra in an effort to make sure that I had not committed some sign error. I proceeded to verify that my results the previous evening were in fact correct, although I had no physical understanding of why the cancellations had occurred.

Aided by many more calculations en route to San Francisco and in subsequent days, I finally realized what was going on. The hypercharge model of BP and BCL had assumed that the field was spatially constant over the size of the experiment, which would be the case if the field was of cosmological origin. However, I had been calculating the effects of a field which could vary spatially over the dimensions of the experimental system. As seen in the rest frame of the K^0–$\overline{K}^0$ system, which is the frame in which the data are typically analyzed, the kaons would see a spatially (and temporally) varying field, and this variation produced an additional γ-dependence which offset the γ^2 dependence arising from the vectorial nature in the field. The shorter the range of this field the greater the γ-dependence, and in the limit of a very short-range field described by a delta function, these two γ-dependences exactly canceled, thus eliminating the criticism of the hypercharge mechanism as an explanation of the observed CP-violation. This observation eventually made it into the invited talk I gave at the 1986 High Energy Conference at Berkeley (Fischbach et al. 1987). For a vector field A_μ with components $[A = 0, A_0 = \sigma\delta(z)]$, which crudely simulates the effects of a short-range potential ΔV, then if the lab (x) and kaon (x') coordinate systems coincide at $t = t' = 0$, then for a boost in the z-direction the potential fA_0' seen by the kaons in their frame is given by Fischbach et al. (1987)

$$fA_0' = \gamma f\sigma\delta(z) = \gamma f\sigma\delta(\gamma\beta t') \approx f\sigma\delta(t') \,, \qquad (6.11)$$

where we assume that $\beta = v/c \approx 1$ in the last step, as is appropriate for high-energy kaons. We see from (6.11) that for a potential of zero range the two sources of γ-dependence exactly offset each other, so that the potential experienced by a high-energy kaon in its rest frame is actually independent of γ.

This result had a significant influence on my thinking, since it revived the possibility that an external hypercharge field could explain both CP violation and the anomalous energy dependence we had found in the high-energy kaon data at Fermilab. As we noted in the published write up of the Berkeley talk, as the range of a putative hypercharge interaction decreases, the γ-dependence of the kaon parameters, such as η_{+-}, ϕ_{+-}, $\Delta m(K_{L,S})$ and τ_S, become "softer", possibly more in line with the gentler γ-dependence that we already reported. When we later became aware of the anomalous geophysical results from Stacey and Tuck, it thus became more plausible that a common mechanism could explain both anomalies.

6.2   Reanalysis of the EPF Experiment

As noted above, shortly after arriving in Seattle, I returned to the question of studying the implications of existing data on possible new long-range forces.

6.2.1   The Review of Gibbons and Whiting

Among the papers that had the most direct influence on our original PRL were those by Stacey and Tuck on the geophysical determination of the Newtonian gravitational constant (Stacey 1978, 1983, 1984, 1990; Stacey et al. 1981, 1986, 1987a,b,c, 1988; Stacey and Tuck 1981, 1984, 1988), and by Lee and Yang on the implications of a long-range coupling to baryon number (Lee and Yang 1955). Additionally, the review by Gibbons and Whiting (GW) in *Nature* (Gibbons and Whiting 1981) played an important role by organizing the then-existing constraints on the strength α and range λ of a putative new long-range force into the now familiar α–λ plot. Among the other experimental results, the GW α–λ plot included both those of Dan Long (1976, 1980) which claimed a deviation from Newtonian gravity, and the results of Riley Newman's group (Spero et al. 1980) which found no discrepancy. A subsequent experiment by Newman's group (Hoskins et al. 1985) further strengthened the limits on non-Newtonian gravity over laboratory distance scales, and these generate the limit labeled "Laboratory" in Fig. 6.9 below.

However, what is of interest from a historical point of view is that the GW review did not include any constraints on α and λ arising from the EPF experiment, or from the subsequent RKD (Roll et al. 1964) or BP (Braginskii and Panov 1972) versions, as we have already noted. Although not explicitly stated by GW, this omission was presumably due to the recognition that for these experiments α would depend explicitly on the composition of the samples. Specifically, for a long-range force arising from a coupling to baryon number B, α would be given by

$$\alpha = -\left(\frac{B_1}{\mu_1}\right)\left(\frac{B_2}{\mu_2}\right)\xi_B \tag{6.12}$$

where $B_{1,2}$ are the baryon numbers of the interacting objects, and $\mu_{1,2}$ the corresponding masses in units of the $^1H_1$ mass (see Appendix 1). In this picture ξ_B is the universal constant which, for composition-dependent experiments, plays the same role as α for composition-independent experiments. Evidently, an analogous equation would apply if the putative long-range force coupled to lepton number (L) or isospin (I), and hence each of these possibilities would generate different constraints on the corresponding constants ξ_L and ξ_I.

As is clear from the above discussion, the phenomenology of composition-independent experiments is qualitatively different from that of composition-dependent experiments, as we explore in more detail in Appendix 1. Had the

GW review been extended to include composition-dependent experiments, the implications of the EPF experiment might have been considered earlier.

6.2.2   Description of the EPF Experiment

The EPF experiment can be thought of as a descendent of the Guyòt experiment, which is in turn a descendent of the Newton pendulum experiment as described in (Fischbach and Talmadge 1999, p. 124). The purpose of Newton's experiment was to search for a possible difference between the inertial mass m_I of an object and its gravitational mass m_G, when the object is suspended from a fiber of length ℓ in the Earth's gravitational field. If θ denotes the angular displacement of the fiber from the vertical, the differential equation describing its motion is

$$m_I \ell \frac{d^2\theta}{dt^2} + m_G g \sin\theta = 0 . \qquad (6.13)$$

For small displacements the oscillation period T is then given by

$$T \approx 2\pi \sqrt{\frac{\ell}{(1+\kappa)g}} , \qquad (6.14)$$

where $m_G/m_I \equiv 1 + \kappa$. By comparing the periods T_1 and T_2 of two masses of different composition Newton was able to set a limit on $\Delta\kappa_{1-2} \equiv \kappa_1 - \kappa_2$ from

$$\Delta\kappa_{1-2} \approx -\frac{2(T_1 - T_2)}{T} . \qquad (6.15)$$

Newton found $|\Delta\kappa| \lesssim 1/1000$, a result which was later improved upon by Bessel who obtained $|\Delta\kappa| \lesssim 1/60{,}000$. In the Guyòt experiment the normal to the surface of a pool of mercury was compared to the normal of masses of different composition suspended over the mercury. Note that all of these experiments utilize objects suspended from fibers, and variants of this technology continue to the present as the source of the most sensitive limits on $\Delta\kappa$.

In the EPF experiment several balances were used, one of which is depicted in Fig. 6.2. What will be particularly relevant in the ensuing discussion are these features: the triple-layer walls for thermal protection, and the thermometers riveted to the apparatus, which attest to the concern of EPF about thermal influences. Additionally, the sample to be tested and the Pt standard are located at different elevations in the Earth's gravitational field, making this apparatus particularly sensitive to gravity gradients. EPF corrected for gravity gradients by taking various differences and ratios of their measured quantities.

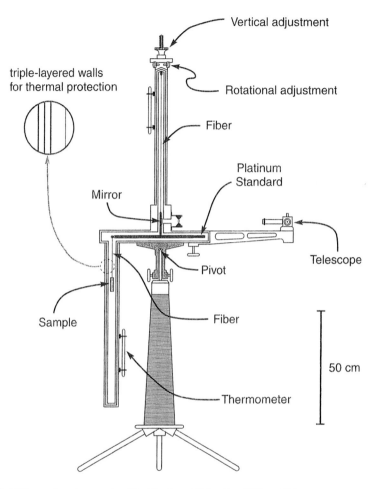

Fig. 6.2 EPF experiment apparatus (Fischbach and Talmadge 1999, p. 133)

6.2.3   Evaluation of B/μ for the EPF Samples

Late in September of 1985 Carrick and I sat down to evaluate the baryon number-to-mass values B/μ for the EPF samples. At this point we were using the data EPF compiled in the table on p. 65 of their paper, in which the accelerations of various test masses were compared to those of a Pt standard. With my limited knowledge of German I knew enough to discern what the samples were, but not enough to recognize at that time that these were not the actual raw data that EPF had measured (see below). For copper, water, and magnalium (a magnesium–aluminum alloy) the compositions were well known, and hence it was straightforward to calculate the corresponding B/μ. Since I had done such calculations in connection with my previously discussed paper connecting Lorentz invariance and the EPF experiment

(Fischbach et al. 1985), Carrick had no problem understanding my explanation of what to do. At that point, I left the calculations to Carrick, and took off with my family for a weekend of hiking in the mountains.

6.2.3.1   The Copper Sulfate Datum

By Monday, Carrick had analyzed three of the EPF data points. Surprisingly, when the results for the acceleration difference in each pair of samples ($\Delta\kappa$ in the EPF notation) were plotted against the difference in the baryon number-to-mass ratio [$\Delta(B/\mu)$ in our notation], the three points fell along a common sloping line, as would be expected if there did in fact exist a new long-range force whose source was baryon number or hypercharge. Of course, this was hardly compelling evidence for a new force, particularly since the data (and associated errors) that we were using were those presented by EPF in their table on p. 65 of their paper, and had large uncertainties. As I shall discuss below, the error bars on their data were artificially large, which made it rather more likely that a satisfactory fit could be obtained with three points.

We next agreed to analyze the copper sulfate datum. Carrick returned to his office, but when he reappeared in mine he was clearly dejected. The copper sulfate datum did not fall along the line determined by the previous three points, and the best fit to what were now four points was no longer even minimally suggestive of anything interesting. Even though we had no "right" to be despondent, we both clearly felt a sense of loss. (I remember thinking at the time of the biblical story of Jonah and the shade tree.) Although Carrick was always extremely careful, and rarely made even small mistakes, I felt obliged to go over his calculation just to make sure he had not slipped up. We began with me asking him what the chemical formula was for copper sulfate, and he told me (correctly) $CuSO_4$. As a high school student I had become fascinated with chemistry, and entered Columbia in 1959 as a chemistry major. No sooner had Carrick told me the formula he used for copper sulfate, I   recalled that the familiar blue crystals that we associate with copper sulfate contain water of hydration. As would be both poetic and prophetic for what would become known as the fifth force, I guessed that the blue crystals existed in the pentahydrate form, $CuSO_4 \cdot 5H_2O$.

My interest in chemistry had been sparked in part by my uncle William Spindel, who had been at various times a professor at Rutgers University and Yeshivah University. For my 15th birthday he rewarded my interest in chemistry with a gift of the 38th (1956–1957) edition of the CRC Handbook of Chemistry and Physics, and it was with me during my sabbatical at the University of Washington (UW). I reached for it and turned to page 516, and there it was: the blue triclinic crystals were indeed $CuSO_4 \cdot 5H_2O$. I asked Carrick to go back and recalculate the copper sulfate datum assuming that the sample was in fact the blue crystals. He returned about an hour later beaming: using the correct formula, $CuSO_4 \cdot 5H_2O$ now fit beautifully on the same straight line determined by the previous three points. As I looked at his graph I felt an adrenaline rush which was my body's way of telling

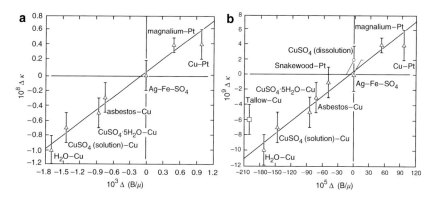

Fig. 6.3 Dependence of Eötvös parameter on baryon number: (**a**) is from Fischbach et al. (1986a) and (**b**) is Fig. 2 of Fischbach et al. (1988)

me that we were seeing an interesting effect. From that point on I felt convinced that the remaining EPF data would fall along the same line, and they did (see Fig. 6.3).

In hindsight Carrick and I were lucky that the copper sulfate datum was the 4th to be analyzed, and not among the first or last three. Had it been among the first three there would have been at the outset no obvious pattern, and we might have quit the analysis of the EPF paper at that point. Had it been among the last three, by which time a pattern would have been evident, we might simply have viewed the (incorrect) result obtained as an outlier, and not bothered to establish its correct formula. But having the correct formula for copper sulfate was important because it led to the recognition that, surprisingly, platinum and copper sulfate had very nearly identical B/μ values, although they differ in every other known physical attribute. Interestingly, the EPF data show that they have very nearly the same acceleration in the Earth's gravitational field. Is this an extraordinary coincidence, or perhaps another hint of a new interaction? The significance of this observation will be discussed in Sect. 6.2.6.

Although EPF explicitly state that they used "crystallized copper sulfate" (Szabó 1998, p. 2), we did not have the translation available to us at that time, and hence the form of copper sulfate remained an issue for us until we resolved it to our satisfaction as described below.

With some help from colleagues at UW we decided to show that even if EPF had started with the anhydrous form of $CuSO_4$, which is a whitish powder, that in the course of their experiment they would have ended up with $CuSO_4 \cdot 5H_2O$ due to absorption of water from the atmosphere. We began by heating a sample of blue crystals for several hours to drive out the water, and then literally ran to another room to weigh the sample. Running was necessary since this was a rainy period in Seattle, and the ambient humidity was sufficiently high that the sample started to turn blue immediately while we were en route to weighing it. We repeatedly weighed the sample over the next few weeks, and found that the sample—initially $CuSO_4$—rapidly absorbed water, and asymptotically approached a composition $CuSO_4 \cdot 4.7H_2O$. Had EPF actually started with $CuSO_4$ rather than with $CuSO_4 \cdot 5H_2O$, they would have found their sample mass increasing in time, which would have thwarted their attempt to accurately measure the acceleration of this sample.

6.2.3.2 Other EPF Samples

We next turned our attention to snakewood, which is an exotic dense wood whose uses include violin bows and other musical instruments (Fischbach et al. 1988).[3] We succeeded in obtaining samples of snakewood from a local instrument maker, Alex Eppler, and confirmed that they were in fact snakewood through the U.S. Forest Products Laboratory. My hosts, the Institute for Nuclear Theory at UW generously agreed to underwrite the cost of a chemical analysis of snakewood, and when the results of this analysis were used to compute B/μ for snakewood we found a surprise: notwithstanding the obvious physical difference between snakewood and more familiar woods, the resulting value of B/μ was virtually identical to that of its main component, cellulose $[(C_6H_{10}O_5)_x]$. Moreover, this would be true for all of the woods we analyzed (Fischbach et al. 1988, Table IX). Carrick thought that it would be amusing to connect the disciplines of forestry (trees) and quantum physics (B/μ) by compiling B/μ for 20 types of wood. This table made it into his Ph.D. thesis, and (to my great surprise and his delight) got into our summary paper in *Annals of Physics* published in 1988 (Fischbach et al. 1988).

The last sample we addressed was *talg* (tallow, fat, suet, ...) whose composition could vary widely depending on (among other issues) its water content. (When I visited Stanford on November 13, 1986 to give a talk about our paper, Bill Fairbank noted that Dicke had erroneously translated *talg* as talc, which is actually *talk* in German.) The best we could do was to estimate B/μ for typical animal fat, and not surprisingly, this datum appears as somewhat of an outlier on the line determined by the other samples.

6.2.3.3 The Ag–Fe–SO₄ Datum

Among the pairs of materials whose accelerations were compared by EPF were the reactants before and after the chemical reaction

$$Ag_2SO_4 + 2FeSO_4 \longrightarrow 2\,Ag + Fe_2\,(SO_4)_3. \tag{6.16}$$

---

[3] The 2003 Summer catalog from Fahrney's in Washington, D.C., featured the Faber-Castell 2003 Pen-of-the-Year crafted in snakewood, which it characterized as "a beautiful and costly wood often used for violin bows and works of art." The pen was priced at $790.

EPF noted that their interest in this process was motivated by an earlier paper in which Landolt suggested the presence of some anomaly. At first glance this datum would seem uninteresting in the present context, since the chemical constituents before and after the reaction are evidently identical. Thus it would seem unsurprising that EPF found $\Delta\kappa = (0.0 \pm 0.2) \times 10^{-9}$ for this pair, i.e., the expected null result.

However, there is much that can be learned from this datum as was pointed out to me in a personal communication from Clive Speake. To begin with the Landolt reaction produces Ag which precipitates out of the original solution. Clive estimated that had there been no correction for differences in the centers-of-mass of the reactants, then EPF should have found $\Delta\kappa = +19 \times 10^{-9}$ instead of their published null result quoted above (Fischbach et al. 1988, p. 34). We can infer from Clive's astute observation that EPF clearly understood this problem and must have taken the proper steps to deal with it. This is, after all, not surprising given that Eötvös was arguably the world's leading expert at that time on gravity gradients, and that his torsion balances were specifically designed to measure gravity gradients. Further analysis of this datum can be found in Fischbach et al. (1988), which also discusses the implications of the null result for a possible magnetic influence on the EPF apparatus.

Unfortunately the details of how EPF corrected for either gravity gradients or magnetic effects do not appear in their published paper. As we have noted above, the introduction to the EPF paper states that the current version represents a "considerable abridgement" of the original size of this work. It is reasonable to presume that the original draft, which Eötvös himself prepared, might have included a more detailed discussion of this datum.

The practical impact of this datum in the earliest days following publication of our original work was significant—at least to me. It indicated that EPF must have paid careful attention to a variety of potential problems which could have produced spurious non-null signals, along the lines first suggested by Dicke. My confidence in the validity of the EPF data further increased following my visit to Hungary in 1988, which included a visit to ELGI (the Geophysical Observatory in Tihany) where I had the opportunity to examine some of the Eötvös balances in detail. The sketch on p. 133 of Fischbach and Talmadge (1999) shows the presence of thermometers which were attached to the balance, presumably to mitigate the effects of temperature fluctuations, but were not discussed in the EPF paper. A more detailed discussion of my visit to Hungary is given in Sect. 6.3.7.

6.2.4 Translation of the EPF Paper

The EPF paper was written in German. However, since I know very little German it would have been difficult for me to embark on an analysis of that paper but for the fact that their results were summarized in a convenient table on p. 65 of the original paper Eötvös et al. (1922) and Szabó (1998, p. 295) (see Fig. 6.4). In that table the data are presented in the form of the acceleration differences of the various

	$\kappa - \kappa_{pt}$
Magnálium	$+\,0{,}004\;.\;10^{-6} \pm 0{,}001\;.\;10^{-6}$
Schlangenholz	$-\,0{,}001\;.\;10^{-6} \pm 0{,}002\;.\;10^{-6}$
Kupfer	$-\,0{,}004\;.\;10^{-6} \pm 0{,}002\;.\;10^{-6}$
Wasser	$-\,0{,}006\;.\;10^{-6} \pm 0{,}003\;.\;10^{-6}$
Kristall. Kupfersulfat . . .	$-\,0{,}001\;.\;10^{-6} \pm 0{,}003\;.\;10^{-6}$
Kupfersulfatlösung	$-\,0{,}003\;.\;10^{-6} \pm 0{,}003\;.\;10^{-6}$
Asbest	$+\,0{,}001\;.\;10^{-6} \pm 0{,}003\;.\;10^{-6}$
Talg	$-\,0{,}002\;.\;10^{-6} \pm 0{,}003\;.\;10^{-6}$

Fig. 6.4 Table of results of the EPF experiment taken from p. 65 of Eötvös et al. (1922)

test samples compared to a platinum standard (this is denoted as $\kappa - \kappa_{Pt}$ in their notation). Following our analysis of the $CuSO_4 \cdot 5H_2O$ datum discussed above, the remaining samples did indeed fall along a common straight line. This was obviously an exciting and surprising result, and so I set out to write this up for PRL.

As noted above, it was critical to confirm that EPF were measuring the acceleration differences to the Earth in each pair of materials. This would ensure that the non-null EPF effect would not conflict with the null results from the more sensitive experiments of Roll, Krotkov, and Dicke (RKD) (1964), and that of Bragniskii and Panov (BP) (1972), which compared the accelerations of test samples to the Sun. To this end I enlisted the help of Peter Buck who was a postdoc from Germany at the Institute for Nuclear Theory, where I was. I asked Peter to initially read just enough of the EPF paper to confirm that they were measuring accelerations to the Earth, which he did. This point is noted explicitly on the first page of our PRL (Fischbach et al. 1986a).

As the PRL draft was proceeding I decided one day to page through the EPF paper to see what I could glean from it. Notwithstanding the fact that I could not read German, I was able to discern that there were results in the body of the paper that did not appear in the summary I had been using. Working with Peter Buck, I eventually came to the understanding that the results tabulated on p. 65 of the EPF paper, were not the raw results from their experiment. Interestingly, the results that appeared in the body of the paper were more statistically significant than those appearing in the table, in the sense that the deviations from the expected null results were systematically larger than for the tabulated results. As I discuss below, $(\kappa - \kappa_{Pt})$ for water was $-(6 \pm 3) \times 10^{-9}$, which is a 2 standard deviation (2σ) effect, whereas the original $(\kappa_{water} - \kappa_{Cu})$ datum given on p. 42 of the EPF paper is $-(10 \pm 2) \times 10^{-9}$ which is a 5σ effect.

My "discovery" of the results in the body of the EPF paper made it clear that we had to understand what EPF had actually done in greater detail, and this necessitated translating the entire paper from German into English. Fortunately I was able to assemble a team at the Institute for Nuclear Theory to carry out this task. In addition to Peter Buck, the team consisted of J. Achtzehnter, M. Bickeböller, K. Bräuer and G. Lübeck, aided by Carrick who knew some German. From the translation it became clear that the entries in the table were obtained by combining the actual raw results in the body of the paper in such a way as to infer a comparison of the various samples to Pt (Fischbach et al. 1988, p. 14). Using water as an example the water datum was inferred by writing

$$\kappa_{\text{water}} - \kappa_{\text{Pt}} = (\kappa_{\text{water}} - \kappa_{\text{Cu}}) + (\kappa_{\text{Cu}} - \kappa_{\text{Pt}}) \,, \tag{6.17}$$

which, when numerical values are inserted, gives

$$(-10 \pm 2) \times 10^{-9} + (4 \pm 2) \times 10^{-9} = \left(-6 \pm \sqrt{2^2 + 2^2}\right) \times 10^{-9}$$

$$= (-6 \pm 3) \times 10^{-9} \,. \tag{6.18}$$

As can be seen from this example, the effect of combining their raw data in such a way as to infer a comparison of each sample to Pt reduced the statistical significance of the quoted result. Since this was systematically true for the remaining data points as well, my initial response was to wonder whether the correlation between $\Delta\kappa$ and $\Delta(B/\mu)$ that had emerged from the table was to a large extent an artifact of the inflated uncertainties in the tabulated $(\kappa - \kappa_{\text{Pt}})$ values (Fig. 6.5).

The content of (6.17) and (6.18) was noted in footnote 13 of our original PRL. Although not discussed further at the time, we privately considered the possibility that Pekár and Fekete had presented the data as they did, referenced to Pt, in order to minimize any suggestion of a conflict with the Weak Equivalence Principle (WEP). The WEP was at the heart of Einstein's General Theory of Relativity published in 1915 (Will 1993), and confirmed following Eötvös' death on April 8, 1919 during the solar eclipse of May 29, 1919. It was thus plausible to assume that Pekár and Fekete were responsible for presenting their data as they did on p. 65 of their paper. However, following the publication of our PRL I received a letter from Wilfred Krause in which he attached a letter written by Eötvös around 1908 (since published in Krause 1988). This letter contains essentially the same summary

Materials compared	Page quoted	$10^8 \Delta\kappa$	$10^3 \Delta(B/\mu)$
Cu-Pt	37	$+0.4 \pm 0.2$	$+0.94$
Magnalium-Pt	34	$+0.4 \pm 0.1$	$+0.50$
Ag-Fe-SO$_4$	39	0.0 ± 0.2	0.00
Asbestos-Cu	47	-0.3 ± 0.2	-0.74
CuSo$_4 \cdot$ 5H$_2$O-Cu	44	-0.5 ± 0.2	-0.86
CuSO$_4$(solution)-Cu	45	-0.7 ± 0.2	-1.42
Water-Cu	42	-1.0 ± 0.2	-1.71
Snakewood-Pt	35	-0.1 ± 0.2	?
Tallow-Cu	48	-0.6 ± 0.2	?

Fig. 6.5 Table from Fischbach et al. (1986a)

of the EPF data as would later appear in the published EPF paper. As Krause notes
"...the idea of referencing all data to platinum was familiar to Eötvös, and not
introduced after his death by Pekár and Fekete." Krause speculates that "...Eötvös
planned new measurements under conditions of reduced man-made mechanical
noise, an undertaking which eventually had been hampered by World War I." These
planned new investigations are in fact referred to at the beginning of the EPF paper.
However, as we discuss below, to the best of our knowledge the correlation between
their measured values of $\Delta\kappa$ and the non-classical quantities $\Delta(B/\mu)$, cannot be
accounted for by any classical effect such as "mechanical noise".

Armed with our translation Carrick and I went through the EPF paper and
replotted their results using the data presented in the body of the paper. Happily,
the effect of using the original data to plot $\Delta\kappa$ versus $\Delta(B/\mu)$ was to increase the
statistical significance of the slope in this plot to 8σ, which was a dramatic non-null
result. To ensure that readers of our paper who were interested in reproducing our
plot used the correct data, we decided to cite in Table I of our paper the page in the
original EPF paper where each datum was listed.

In 1998, which was the 150th anniversary of the birth of Eötvös (July 27, 1848),
the Eötvös Roland Geophysical Institute (ELGI) of Hungary published a volume
entitled "Three Fundamental Papers of Roland Eötvös", one of which was the EPF
paper, and we were invited to contribute our translation to this volume, which
was published along with the original German paper (Szabó 1998). Carrick and
I revisited our original translation, with the goal of making it more readable to
modern researchers while at the same time adhering as closely as possible to the
original text. Significantly, this translation corrects a number of typographical errors
in the original EPF paper. These were uncovered by Carrick who carefully checked
their final results against the raw torsion balance data presented by EPF. These
corrections are identified in various footnotes in the text of the translation, and are
distinguished from the footnotes present in the original EPF paper.

6.2.5   The Refereeing Process

Our paper was received by PRL on November 7, 1985. At that time the leading
experts in the world on the Eötvös experiment were Robert Dicke at Princeton and
V.B. Braginskii at Moscow State University. It was thus natural to assume that Dicke
would be one of the referees, and he was. Normally the referees at PRL (and at
most other physics journals) are anonymous, but Dicke chose to identify himself
through a message he sent directly to me on November 20 (see Appendix 3). In that
message he raised the possibility that the EPF data could be explained in terms of
conventional physics, and asked us to reanalyze the EPF data to test his suggestion.
Specifically, Dicke began by noting that the brass containers in which the EPF
samples were contained were of different lengths, and hence had different cross-
sectional areas. Thus if there were a thermal gradient present in the vicinity of the
EPF apparatus there could arise an air current, and this could lead to a differential

force on the two samples being compared in each pair. Given that the various samples used by EPF had very nearly the same masses, it follows that samples of higher density were contained in cylinders of smaller volume and hence of smaller surface area, owing to the fact that they had similar diameters (Fischbach et al. 1988, p. 48)). The Dicke model, later elaborated upon by Chu and Dicke (1986), provided a nice pedagogical example of how a purely conventional mechanism could have produced a differential signal in the EPF experiment which depended on a property of the samples, specifically $1/\rho$, where ρ is the sample density (Fischbach et al. 1988, p. 49).

Dicke's message to us was gracious and indicated that he was inclined to accept our paper once we addressed his question. Carrick and I set about immediately to analyze Dicke's model. Leaving aside the details of exactly how such a mechanism might work, which are discussed in detail in Fischbach et al. (1988), the simple question at that time was whether any such correlation actually existed. Carrick plotted the data, which are exhibited in Fig. 7 of Fischbach et al. (1988) (see Fig. 6.6). It was immediately clear that the fit was quite poor, with the snakewood–Pt datum falling far off the best-fit line. We conveyed this result to Dicke on November 27 (Appendix 3), and eventually suggested that a note be added to our paper presenting this result. He agreed, and recommended to the PRL editors to allow us to include such a note. The editors agreed even though its inclusion would

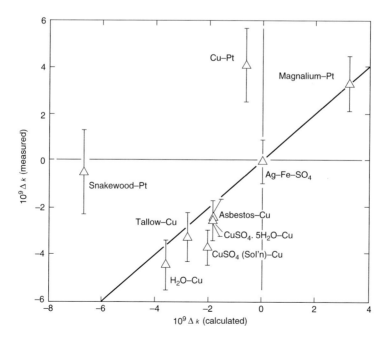

Fig. 6.6 Figure 7 from Fischbach et al. (1988)

lengthen our paper beyond the maximum allowed by PRL at that time.[4] In that note we observed that the failure of this model, in contrast to one based on B/μ as the charge, was

> [...] a consequence of two special properties of B/μ: (6.1) it has an anomalously low value for hydrogen, and (6.2) it has a maximum near Fe and is lower toward either end of the Periodic Table.

As noted above, the question raised by Dicke was later elaborated upon in a Comment published in PRL (Chu and Dicke 1986), to which we responded in Fischbach et al. (1986b). Surprisingly, this exchange of short comments was picked up by the *New York Times* in a story "Physicists Challenge Theory of a 'Fifth Force' beyond Gravity," by John Noble Wilford that appeared on October 18, 1986.

Considering the fact that our PRL was suggesting the presence of a new force in nature it may seem surprising that the refereeing process went as smoothly as it did.[5] I would identify three likely reasons for this. Most significantly, our reanalysis of the EPF experiment did not challenge the work of anyone who was still alive. In fact the only earlier work which our PRL may have called in question was that of Renner (1935), which had been previously criticized by Dicke (1961) and Roll et al. (1964). Furthermore, we took pains to note in our paper that the experiment of Roll, Krotkov, and Dicke (RKD) (1964), and that of Braginskii and Panov (BP) (1972), would not have been sensitive to a new force whose range was of order 1 km, since both of these experiments measured the accelerations of pairs of materials to the Sun. Hence any evidence arising from our reanalysis of the EPF experiment suggesting a new intermediate range force would not contradict the more precise RKD and BP experiments.

The second feature of our original PRL paper, which may have aided its rapid acceptance, was the recognition that various theories predicted the existence of new long- or intermediate-range forces. As we have noted previously, our original PRL paper was motivated in part by the elegant 1955 paper by Lee and Yang (1955), who used the EPF paper to set limits on a long-range force coupling to baryon number. Additionally, one of our primary motivations was the geophysical determination of the Newtonian gravitational constant G by Stacey and Tuck (1981) and Holding and Tuck (1984) which had been motivated in turn by an elegant and prescient paper by Fujii dealing with modifications of Newtonian gravity (Fujii 1971). In recent years theories based on supergravity, supersymmetry, and string theory have produced many candidates for new macroscopic fields, which explains in part the continuing

---

[4]In contrast, when a similar situation arose with respect to a story about our work in *National Geographic*, the editors insisted that their word count limit be strictly enforced, as discussed in Sect. 6.4.3.

[5]One measure of this surprise is a published comment from Lawrence Krauss, then a young assistant professor at Yale (Krauss 2008): "I reacted with surprise that the paper [our PRL] had survived the refereeing process, which at the time had very strict self-imposed requirements of general interest, importance, and validity." See also Sect. 6.4.2.

interest in fifth force tests, specifically, tests of both the weak equivalence principle
and the gravitational inverse square law.

The third factor which contributed to the relatively smooth referee process was
the fortunate choice of reviewers. As noted above, Robert Dicke, the towering figure
in the field, was both insightful and gracious, and his recommendation to publish our
paper no doubt carried great weight with the editors. At that time I did not know who
the second referee was. Only later did I learn from Vern Sandberg (who had been
at Los Alamos at the time) that he was the second referee. Vern and I have had
several conversations about our paper, which he clearly read quite carefully. He is
by all accounts a very conscientious reviewer, and he also shares my view of the
refereeing process. In my case it is derived in part from a conversation I overheard
as a young faculty member in which Francis Low of MIT said something to the
following effect to a colleague: when reviewing papers he gives authors the benefit
doubt, because publication is cheap, but not on grant proposals because the available
pot of money is limited. The actual reports from Dicke and Sandberg are given in
Appendix 4, along with the correspondence with PRL.

6.2.6   An Alternative Explanation

As noted above, one of the arguments against an explanation of the EPF results as
an "environmental" effect, as had been proposed by Dicke (see Sect. 6.2.5), was the
fact that the EPF correlation depended on the value of B/μ for each sample and
this was a non-classical parameter. One way of expressing the implication of this
fact is the observation that two of the materials employed by EPF were Pt ($B/\mu =$
1.00801), and $CuSO_4 \cdot 5H_2O$ ($B/\mu = 1.00809$) which were very nearly equal. There
is no conventional physical quantity (e.g. density, electrical conductivity, etc.) which
is the same for these two materials. By combining the EPF data for Pt–Cu and
$CuSO_4 \cdot 5H_2O$–Cu, we can find (Fischbach et al. 1988)

$$\frac{\Delta(B/\mu)_{\text{Cu-Pt}}}{\Delta(B/\mu)_{\text{CuSO}_4 \cdot 5\text{H}_2\text{O-Cu}}} = \frac{+94.2 \times 10^{-5}}{-85.7 \times 10^{-5}} = -1.10 \,, \tag{6.19}$$

$$\frac{\Delta\kappa_{\text{Cu-Pt}}}{\Delta\kappa_{\text{CuSO}_4 \cdot 5\text{H}_2\text{O-Cu}}} = \frac{(+4.08 \pm 1.58) \times 10^{-9}}{(-4.03 \pm 1.33) \times 10^{-9}} = -1.01 \pm 0.51 \,. \tag{6.20}$$

The close agreement between the measured $\Delta\kappa$ ratios, and the theoretically
expected values based on the $\Delta(B/\mu)$ ratios, appears to provide strong support to the
view that EPF were seeing an unconventional effect uniquely tied to a non-classical
quantity such as baryon number or hypercharge. (We recall that baryon number and
hypercharge were only introduced into the physics literature many years following
publication of the EPF paper.)

To our great surprise this conclusion would be challenged by a 1991 paper that
Carrick received from PRL to review. The authors were Andrew Hall and Horst

Armbruster who were then, respectively, a graduate student and faculty member at Virginia Commonwealth University in Richmond, Virginia. The primary driving force (and first author) was Hall, who was claiming in this paper that he had constructed a phenomenological "charge" which could explain the EPF data just as well as our hypercharge hypothesis. This "charge" Q depended on the intrinsic nuclear spins of the EPF samples and was defined by

$$Q = M\delta , \qquad \delta = \begin{cases} 1 & \text{if } J > 0 , \\ 0 & \text{if } J = 0 , \end{cases} \qquad (6.21)$$

where M is the mass of the nucleus, and J is its nuclear spin (in units of $\hbar$).

Carrick and I greeted the Hall/Armbruster (H/A) paper with a great deal of skepticism. We were no doubt biased in our view that B/μ was not only the correct "charge" to explain the EPF data, but that it was also unique by virtue of the preceding discussion. Additionally, we could not understand how a "charge" which depended on nuclear spin could be relevant in an experiment utilizing samples which were unpolarized, as was presumably the case for the EPF samples. Nonetheless we were determined to take this paper seriously, and so we decided to verify Hall's claim that Q given by (6.21) could in fact explain the EPF data.

As it turned out I had a dinner engagement the day Carrick received the paper, but I arranged with him to return to his office around 10 PM, at which time we would then work on the H/A paper as long as needed. When I returned we divided the work as follows: Carrick would modify his existing code to allow us to compute Q for the EPF samples. While he was doing that I busied myself with the task of determining the nuclear spins of the elements in the EPF samples from various tables. By midnight we were able to compute the analog of our plot of $\Delta\kappa$ versus $\Delta(B/\mu)$, where $\Delta(B/\mu)$ was now replaced by ΔQ for each pair of samples. Carrick hit the ENTER key on his NeXT computer, and instantly a figure appeared on his screen which looked almost indistinguishable from our published figure (Fig. 6.7). Although the relative positions of the various data points were different, the overall quality of the fit was as good as ours using $\Delta(B/\mu)$.

It would be difficult to overestimate the significance of the H/A paper, had it turned out to be correct. The design of any experiment can depend critically on the specific theory being tested. For example, to test the B/μ theory we had advanced in our original paper, it was advantageous to compare samples widely separated in the periodic table, such as Al–Au, Al–Pt, Be–Cu, and so on. For the purpose of repeating the EPF experiment, the nuclear spins of the sample would be irrelevant in a B/μ picture, whereas they would evidently have been relevant in the actual EPF experiment in the H/A framework. The fact that experiments were framed in terms of specific theories is a recurring theme in the history of the fifth force, as we shall see.

Carrick (and I) accepted the H/A paper for publication in PRL. However, their paper never appeared in PRL, presumably because it must have been rejected by another referee. (Under the policy followed by PRL—at least at that time—a split

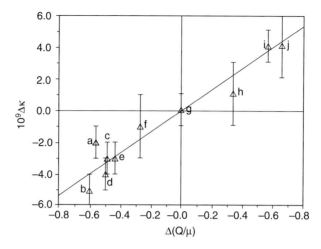

Fig. 6.7 Plot of $\Delta\kappa$ versus $\Delta(Q/\mu)$, where Q is the "charge" defined in (6.21). This plot is Fig. 8.1 from Fischbach and Talmadge (1999), and the labels on the samples are defined in Hall (1991)

decision was typically resolved against the authors.) Eventually I contacted Andrew Hall and informed him that Carrick and I had reviewed his paper (positively) for PRL. He then confirmed that another referee had rejected his paper. Since Carrick and I felt that the H/A results should be publicized, we arranged to include a revised version of this paper as my contribution to a conference in Taiwan (Hall 1991), which was co-authored by Horst Armbruster and Carrick.

Some years later I learned who the other reviewer of the H/A paper was. Not surprisingly, the shortcomings of the original H/A paper which necessitated the revisions that Carrick and I felt should be incorporated into (Hall 1991), also concerned this reviewer, and formed the basis for rejecting the H/A paper.

The story of the Hall "spin-charge" raises the broader and deeper question of the reproducibility of experiments, a subject which has been much in the news recently.[6] As we have noted above, the design of any experiment to search for the presence of a fifth force depends to a great extent on having some model of how the sought-after effect depends on whatever aspects of the experiment are under the control of the experimentalist. This might include the choice and preparation of samples, design of apparatus, data analysis, etc. In fact the very notion of repeating an experiment carries with it some notion that the effect being studied should not depend in a significant way on when the original and subsequent experiments were carried out, which may not always be the case.

---

[6]See *New York Times*, Sunday Review, February 2, 2014, p. 12. See also Centerforopenscience.org.

6.3   Immediate Aftermath of Publication

As noted previously, our paper was published in PRL on January 6, 1986. By coincidence it was the first paper in PRL published in 1986, although I doubt that this had much to do with the attention it was about to receive. Our Christmas vacation had been delayed due to an unusually heavy smog that settled over the Seattle area during the Christmas period, which affected air travel among other inconveniences.

As a consequence of the smog, and the unpleasant weather we encountered in California, I was suffering from a massive head cold by the time we left California for home on Sunday, January 5, 1986. By the time we landed in Seattle I was experiencing a significant hearing loss resulting from the congestion associated with the cold, along with a persistent cough. When I arrived in work the next day both the hearing problem and cough had improved, but only slightly. And so when the phone rang in my office at around 11 AM on Monday, I wasn't quite sure that I was hearing properly when John Noble Wilford from the New York Times called to talk about our paper—which I had yet to see in print.

My conversation with John was very pleasant, although he was a little vague when I asked the obvious question, how he even knew about our work. I gathered from what he did say that he had a number of contacts who would suggest stories to him. By Tuesday, January 7, I had been sent a sketch of the alleged Galileo experiment on the leaning tower of Pisa, which would appear the next day with the full story. By Tuesday evening there was a brief mention of our work on the CBS-TV evening news, anchored by Bob Schieffer, and somewhat longer story on NBC radio.

The headline on John Noble Wilford's story on Wednesday, January 8, "Hints of 5th Force in Universe Challenge Galileo's Findings", introduced the notion of a "fifth force". In this reckoning the other four forces, in order of decreasing strengths, are the strong, electromagnetic, weak, and gravitational. Although some might quibble with drawing a distinction between the electromagnetic and weak manifestations of what we now consider to be the unified electroweak interaction, the notion of a generic "fifth" force has made its way into the published literature usually without attribution. As used, this refers to a long-range non-gravitational force presumably arising from the exchange of any of the ultra-light quanta whose existence is predicted by various unification theories such as supersymmetry. Although I cannot be sure of historical precedents, this is likely to be a rare (and possibly unique) instance in which a widely used physics concept owes its name to a journalist.

Wilford's story appeared Wednesday January 8, surprisingly on the front page, along with the aforementioned picture. My day began, unfortunately, at approximately 4 AM with a call from an Australian reporter who was unaware of what time zone Seattle was in. He was interested in the connection between our paper and the work of his fellow Australians Frank Stacey and Gary Tuck, which we had cited as part of the motivation for our work. After I politely indicated to him what time it was for me, we agreed to have a longer talk later in the day, which we did. After breakfast I drove to my office, stopping along the way at the UW bookstore to pick up a half-dozen copies of the *Times*. By the time I reached my office I found a stack of phone messages from reporters on my desk, and for the remainder of the day I did nothing but try to respond to these, while at the same time answering calls as they came in. Additionally reporters from local Seattle media showed up at my door, and I was eventually forced to unplug my phone in order to make time available for them.

Some time after 6 PM I decided that it was time for dinner, given that I had nothing for lunch, and so I left for our rental home in Bellevue. Ordinarily the traffic on the 520 Floating Bridge across Lake Washington, which connects Seattle and Bellevue, was bothersome. However, given the stressful day that I was now escaping from, the traffic was a blessing of sorts. Absent cell phones, which were still many years in the future, I was able to enjoy 45 min of peace and quiet during which nobody could reach me.

As it turned out, my day was not yet over. Shortly after sitting down to dinner the phone rang, and Janie picked it up. "It's *The National Enquirer*," she said, "and they want to talk to you about your work." During the earlier part of the day I had made a special effort to explain to each reporter what our work was about in terms that I felt were appropriate to his/her level of interest and understanding. So how was I now to explain what we had done to a tabloid such as the *Enquirer*? To my relief the caller was actually Bruce Winstein, who is a high-energy experimentalist then at Stanford, and he was interested in the arrangements for my talk the following Monday at Stanford, which had been arranged long before the *N.Y. Times* story. In an odd twist of events, Bruce's seemingly innocuous phone prank led to an unfortunate interaction with Richard Feynman, as I describe in Sect. 6.3.1.

The first public lecture on our paper was at TRIUMF in Vancouver, Canada which had been arranged for the next day Thursday, also long before the publicity generated by the *N.Y. Times* story. Janie and I had just purchased a new Honda Civic, and I was looking forward to breaking it in on the roughly 300 mile round trip to Vancouver. Carrick and I left early in the morning, and after arriving at TRIUMF I was quickly requested to do a radio interview with the Canadian Broadcasting Corporation (CBC). The only problem was that I still had a lingering cough, which the CBC interviewer indicated was causing them problems. Somehow I managed to suppress my cough long enough to get through the short interview. The talk itself went very well, which was gratifying, since this was the same talk I was going to give the following Monday at Stanford.

6.3.1 Interaction with Richard Feynman

By Friday January 10 a degree of calm had been restored to me and my family. At around 8 PM the phone rang. Janie was busy cleaning up from dinner, I was busy giving Michael a bath, and so it fell to Jeremy to answer the phone. "Dad,

a Mister Fineman is on the phone ..." I picked up the phone, and without even
formally saying "hello" I said something like "Bruce, stop trying to pull my leg,
I've had a very long week ..." From the other end of the phone came, "...this is
Richard Feynman, I am a theoretical physicist at Caltech..." The fact that the caller
had to identify himself made it certain to me that this was in fact Bruce Winstein
calling again from Stanford (recall, no caller ID in those days!) "Bruce, enough is
enough ..." "This is Richard Feynman, I have a few questions about your recent
paper in PRL." By this point I had become convinced that either this was the best
impersonation of Feynman that I had ever heard, or that "Fineman" was actually
"Feynman".

After obliquely complimenting me for actually reading and analyzing the EPF
paper, he launched into his main criticism. In (9) of our paper we used the EPF data
to determine the quantity $f^2 \epsilon (R/\lambda)$, where f is the unit of hypercharge (analogous
to the electric charge e), assuming that an intermediate-range hypercharge force
was responsible for the non-zero slope seen in the EPF data. Since hypercharge
$Y = B + S$, where B is baryon number and S is strangeness, the hypercharge of any
sample of ordinary matter is simply its baryon number B, the sum of its protons and
neutrons.[7] The function $\epsilon(x)$ is given by

$$\epsilon(x) = \frac{3(1+x)}{x^3} e^{-x}(x \cosh x - \sinh x) , \tag{6.22}$$

and is a "form factor" arising from the integration of an intermediate-range
hypercharge distribution over the Earth, assumed to be a uniform sphere of radius
$R = \lambda \cdot x$. In (9) of our paper we found

$$\left[f^2 \epsilon \left(\frac{R}{\lambda} \right) \right]_{\text{EPF}} = (4.6 \pm 0.6) \times 10^{-42} e^2 , \tag{6.23}$$

where e is the electric charge in Gaussian units. By way of comparison, the value
determined from the geophysical data of Stacey et al. which constituted part of the
original motivation for our paper, was

$$\left[f^2 \epsilon \left(\frac{R}{\lambda} \right) \right]_{\text{geophysical}} = (2.8 \pm 1.5) \times 10^{-43} e^2 , \tag{6.24}$$

I had regarded it as miraculous that two experiments as disparate as EPF and Stacey
et al. agreed within an order of magnitude. However, Feynman viewed the factor 16

---

[7] Ordinary matter is composed exclusively of baryons (and not anti-baryons). It follows that a fifth
force arising from a vector coupling whose source is baryon number or hypercharge would give
rise to a repulsive force between ordinary objects. Since gravity is, in contrast, an attractive force,
a number of stories described our original PRL as providing evidence for "anti-gravity". This in
turn has the consequence that in the falling "coin and feather" comparison, the feather falls faster.
See also Sect. 6.4.2.

discrepancy between these two results as a strong argument against our hypercharge hypothesis as an explanation of the EPF results.

Our conversation ended somewhat better than it had started when I apologized for the manner in which I had answered the phone. However, Feynman remained unconvinced by our analysis, and said so publicly in a letter published on January 25 in the *Los Angeles Times*, which had previously carried a story on our work on January 8 (see Appendix 5). It appears from the letter Feynman sent to the *L.A. Times* that he was motivated to respond to the op-ed piece about our paper entitled "The Wonder of It All," which they had published on January 15. Feynman had been asked what he thought of our theory, and he had responded "Not much." In his follow-up letter, which the *L. A. Times* published on January 25 (and which refers to our phone conversation), he felt the need to elaborate on his quoted remark (see Appendix 5). More interestingly, he apparently also felt the need to explain to me in technical terms the basis for his view. The content of this letter represents a tour de force on Feynman's part, especially considering the fact that he was evidently working from the original EPF paper in German. He begins by focusing on the factor of 16 difference between the results in (6.23) and (6.24), with respect to which he and we had different views. He then considers possible scenarios in which various combinations of α and λ in (6.1) could reconcile the available data, but suggests that this is unlikely.

Feynman's tour de force then follows in which he examines in minute detail the various measurements that EPF carried out. This is a very impressive discussion, which concludes with his comment, "Well, that is the best I can do." I know of no other paper which has analyzed the EPF data in this level of detail, and hence to me Feynman's analysis is all the more remarkable. Given the fact that the fifth force implied by the EPF experiment has not been seen in other experiments, it may be that Feynman's general criticisms were correct, although not necessarily for the specific issues he raised. This question is discussed in greater detail in the epilogue (Sect. 6.8).

Given Feynman's well-deserved reputation in the world of physics and beyond, one might have expected his criticism of our paper to have dealt a fatal blow to our work. However, this proved not to be the case: by the time his letter appeared in print on January 25, a number of groups had recognized that the simplistic model of a uniform spherical Earth acting as a source for a putative hypercharge force was inappropriate for a force whose range was hypothesized to be ∼200 m. In fact we had already noted this explicitly following (10) in our original paper (Sect. 6.3.5). For a force of so short a range, local inhomogeneities such as buildings and basements would play an important role in determining the correct functional form for the expression to be used in place of $\epsilon(x)$ in (6.23) and (6.24). As we discuss in Sect. 6.3.5, the recognition of the importance of local inhomogeneities served to clarify both the magnitude and sign of the putative hypercharge force.

The significance of local inhomogeneities led to several papers which were submitted at nearly the same time to PRL, including one by our group (Bizzeti 1986; Milgrom 1986; Thieberger 1986). The submission of our paper was slightly delayed owing to our desire to obtain the approximate dimensions of the building in which it was presumed that EPF carried out their experiment, which we received from Judit Németh (Talmadge et al. 1986, p. 237). In the end we demonstrated that (Talmadge et al. 1986, p. 236) "neither the magnitude nor the sign of the effective hypercharge coupling can be extracted unambiguously from the EPF data without a more detailed knowledge of the local matter distribution." Although our paper was accepted by the reviewers for publication in PRL, in an unusual move the editors of PRL declined to publish any but the first paper to have been received, which was an elegant paper by Peter Thieberger from Brookhaven National Laboratory (Thieberger 1986).

The appearance of the papers on the influence of the local matter distributions, even in preprint form, served to mute Feynman's criticism which in the end appears to have had little lasting impact. What impact it did have was further muted by the Challenger disaster three days later on January 28, 1986, in whose subsequent investigation Feynman played so crucial a role. I do not know whether Feynman was aware of the above papers. However, following the conclusion of the Challenger investigation, in which Feynman famously pointed to the problem with the O-ring seals (by dipping one in ice water), I re-engaged with him on the question of local inhomogeneities through a letter I sent on April 14 (see Appendix 5).

6.3.2 The Talk at Stanford

This was the second public presentation of our paper and, as I anticipated, was more probing. Although Stanford was happy to pay for me to fly from Seattle to San Francisco, I opted to drive instead with Carrick in my new Honda. I had arranged to stay with my close friends Jim and Marilyn Brittingham in Livermore, California where Jim (since deceased) was on the staff of Lawrence Livermore National Laboratory. Carrick and I left Seattle around 7 AM and arrived in Livermore some time between 9 and 10 PM.

The next morning we drove to Stanford, and joined some faculty for lunch. There I met Bill Fairbank for what would prove to be the first of a number of subsequent pleasant encounters. As I noted above, Bill began by complimenting Carrick and me for correctly identifying *talg* as fat or suet. (Credit for this goes directly to Carrick!) At the talk itself the questions were polite, as illustrated by the following from Bruce Winstein. He noted that if we had plotted the EPF result for (Pt–magnalium) rather than for (magnalium–Pt) as we did, that datum would have ended up in the 3rd quadrant of our PRL Fig. 1, rather than in the first, and the figure would have looked less dramatic. I responded by first acknowledging that this would be so, but then noting that this (arbitrary) shift would merely change the "optics" of the figure but not the slope of the resulting line nor its $\sim 8\sigma$ significance, which were the physically important results. I then added that in writing this paper we had included the following sentence specifically to address questions of the sort that Bruce had raised: "Table I gives $\Delta\kappa$ for each of the nine pairs of materials measured by EPF, *exactly as their result is quoted on the indicated page of Ref. 6*" (emphasis added).

By the end of the talk I felt that it had gone sufficiently well that the inevitable calls from members of the audience to their colleagues elsewhere would have converged an overall positive tone.

On the return trip to Seattle Carrick and I were joined by Idella Marx, who flew up from Los Angeles to attend my talk at Stanford and then decided to drive home with us. Idella was a science enthusiast who had hired me in 1963 to expose her children to "fun" science. Idella's husband Louis had founded the Marx Toy Company, and she used her resources to indulge her interest and that of her family in science, physics in particular. What neither of us knew as we started out was that she was about to experience one of the great thrills in her life, a surprise meeting with T.D. Lee (see Sect. 6.3.3).

Our otherwise routine trip back to Seattle revealed another surprise for Carrick and me: somehow we got on the subject of the Pentagon papers dealing with the Vietnam war. They were publicly disclosed in 1971 by Daniel Ellsberg who is married to Idella's stepdaughter Patricia Marx. The resulting story of how various missteps by the prosecution which allowed Ellsberg to go free would have been worthy of a Hollywood movie.

We arrived in Seattle late in the evening of January 14, and dropped Carrick off at this apartment. Idella and I then drove to our place in Bellevue, stopping along the way to pick up the latest issues of *Newsweek* and *Time*. Idella had guessed correctly that both would carry stories on the fifth force, and the *Newsweek* version by Sharon Begley (p. 64) was particularly good. Her story began with a bit of word play which I missed, but which other readers caught: "<u>F</u>ew <u>i</u>mages <u>f</u>rom <u>t</u>he <u>h</u>istory of science. . ."

The talks at TRIUMF and Stanford were the first of more than 75 talks that I gave in many countries on the EPF experiment/fifth force between 1986 and 1992 (when I stopped keeping track). In the early days, bcfore the results of new experiments became available, the EPF experiment and our analysis of their data were on occasion the subject of some pointed exchanges during these talks. I dealt with the associated stress by noting to myself that some day when new experimental results became available, I could sit at the back of the room and watch the authors of these experiments focus on one another, and no longer on me and my co-authors. That day came for me on July 6, 1989 when I was attending the GR-12 conference in Boulder, Colorado, the home of the University of Colorado. Just prior to the session on the fifth force I purchased a bag of popcorn and brought it to the conference. There, sitting in the back row, I enjoyed both the popcorn and the excitement of the experimentalists challenging one another and not me.

6.3.3   Meeting with T.D. Lee

As noted earlier, the recognition that the presence of a new long-range (i.e., $r/\lambda \ll 1$) force could be detected by a violation of the Equivalence Principle originated in a beautiful one-page paper by T.D. Lee and C.N. Yang published in *Physical Review* in 1955 (Lee and Yang 1955). Our 1986 paper had extended the work of Lee and Yang in two ways: First, we modified their formalism to allow for this force to have a finite range, unlike gravitational and electromagnetic forces which are believed to extend over an infinite range. Our second, and more important, contribution was to actually plot the EPF data against our theory.

By an extraordinary coincidence, T.D. Lee had been invited to give a series of three public Danz lectures, one of which he delivered on January 15, 1986, just nine days after the publication of our paper in PRL. This had been arranged before I arrived at UW for my sabbatical, and had nothing whatever to do with the publicity surrounding our EPF paper. Notwithstanding the reference to the Lee–Yang paper in our EPF paper in PRL, I suspect that few of my colleagues at UW fully appreciated the deep connections between these two papers. Lee's visit to UW extended over several days, and I arranged to speak with him personally. He obviously knew of our reanalysis the EPF paper and began by congratulating me for it. After some brief discussion of the paper itself, I got around to asking the obvious question: why hadn't he and Yang actually plotted the EPF data, as we had done, instead of assuming as they did that EPF had obtained a null result? I remember Lee chuckling a bit, and then explaining that their one page paper was written at a time when they were deeply involved in other questions, which they regarded as more pressing, such as parity non-conservation in the weak interactions. (Their EPF paper appeared in March 1955, and their Nobel prize-winning paper on parity non-conservation appeared in October 1956.) We can only speculate on how elementary particle physics might have changed had they taken out the time to actually plot the EPF data as we had done. Would this have riveted their attention on the gravitational interaction rather than the weak interaction? And how long would it have taken for them or somebody else to return to parity non-conservation?

During Lee's lecture on January 15, he exhibited some posters he had hand-drawn to accompany his talk. Following his talk I introduced him to Idella Marx who was thrilled to meet Lee. She gently asked whether she could have the posters, and he graciously agreed. This was clearly the highlight of Idella's stay with us.

6.3.4   Some Wrong Papers

The publicity following publication of our paper in PRL led to a flood of comments and criticisms, many of which we received to review. (See PRL editorial comment: *Physical Review Letters* **56**: 2423 (1986).)

Among the papers that arrived in the white-and-green PRL envelopes were several from colleagues whom I personally knew, or at least knew of, which were flawed. Carrick and I carried out a rough triage on all the incoming papers, which some days were arriving at a rate of one or two a day, in contrast to my expected frequency of one every few weeks. Irrespective of what our decision was, Carrick and I worked closely to clearly explain to the authors, editors, and other potential reviewers the basis for our decision. In the end we found that virtually all of our recommendations were followed, so that relatively few incorrect papers made it into the published literature.

With the notable exception of the Thodberg paper, discussed in Sect. 6.3.5 below, which correctly pointed out a sign error in our paper, many of the papers that we received to review contained conceptual errors of one sort or another. A good example is provided by a criticism of our calculation of the B/μ values for our samples that was raised by two senior physicists, one of whom I knew personally. As we note in Fischbach et al. (1988), given the fact that B/μ is close to unity for all substances, it follows that determining $\Delta(B/\mu)$ requires that the values of B/μ for individual elements be calculated to at least six decimal places. For example, B/μ (Mg) $= 1.008453$ and B/μ (Al) $= 1.008515$. To do this the values of B/μ for each isotope of an element, which are known with great precision, must be properly weighted by the relative abundances of these isotopes in the naturally occurring element. These authors then (correctly) note that these abundances are much less well known. (This is due in part to the fact that the abundances can vary from one location to another due to fractionation.) They then argue (incorrectly!) that the uncertainties in these relative abundances would introduce sufficiently large errors in calculating $\Delta(B/\mu)$ as to preclude drawing the conclusions we did in our paper.

This argument, although superficially convincing, is in fact wrong, and led me to reject this paper. What the authors failed to consider is that the values of B/μ for the individual isotopes of an element are so close to one another that it hardly makes a difference what the relative abundances of a given element are. On p. 26 of Fischbach et al. (1988), we illustrate this point quantitatively using as an example the isotopes of Mg, which is a constituent of the magnalium alloy sample used by EPF. There we show explicitly that the actual fractional uncertainty in the calculation of B/μ is approximately 8×10^{-9}, which is completely negligible.

6.3.5   Shortcomings of Our PRL Paper

It is not uncommon in the world of physics for the same idea or observation to occur independently to more than one individual or group at approximately the same time (see, for example, Sect. 6.3.6). Since I myself had experienced this more than once, it was not surprising that when we found the correlation between the EPF data for $\Delta\kappa$ and our calculated values of $\Delta(B/\mu)$, that I started to worry that some individual/group could stumble upon the same observation. In fact my concern was not unreasonable, since the content of the paper was sufficiently straightforward that, following publication of our paper, I learned that it had been assigned as a graduate or undergraduate homework problem by a number of colleagues at various institutions. This self-imposed time pressure resulted in some oversights which, luckily, did not detract from the basic message of the paper.

The most obvious shortcoming was an error we made in the sign of the putative fifth force as inferred from the EPF data. If the force between a source and a test mass is proportional to the product of their respective baryon numbers (or hypercharges), which is what the EPF correlation indicated, then that force had to be intrinsically repulsive since all stable matter has positive baryon number. This leads to clear predictions for the signs of the acceleration differences $\Delta\kappa$ for the various EPF sample pairs. Shortly after our PRL appeared Thodberg (1986) correctly pointed out that in the simple model we were assuming, where a spherical Earth was the source of the observed acceleration differences, the sign of $\Delta\kappa$ between Cu and water as measured by EPF could correspond to an attractive (not repulsive) force.

In the course of writing our paper Carrick had drawn attention to the sign problem, and its connection to both the model of the Earth and the influence of the local matter distribution (see discussion below). My view was that since the sign problem would take some time to sort out, particularly the effects of the local matter distribution, we should not risk the possible consequences of delaying submission of our paper. This view was bolstered by my conviction that the reviewers of our paper would surely require major revisions, which would then allow us the time needed to deal with the sign question. To our surprise our paper was quickly accepted by PRL, with only the minor addition suggested by Dicke, as discussed in Sect. 6.2.5. However, since we clearly appreciated the importance of the local matter distribution, specifically as it would bear on the comparison of (9) and (10) of our paper, we added a note to this effect following (10). What Thodberg's observation pointed out was that understanding the local matter distribution was also necessary to account for the sign of $\Delta\kappa$ for Cu–H_2O as measured by EPF.

To understand how an apparently attractive force can emerge from an interaction which is intrinsically repulsive, imagine that the Earth is a completely uniform sphere, except for a huge hole located somewhere in the vicinity of the EPF experiment. It is then easy to see that the absence of the repulsive force that would have arisen if the hole were not there, would effectively look like the presence of an attractive force in the presence of the hole.[8] To quantify this effect we set out to find the dimensions of the buildings where EPF were presumed to have carried out their experiment. As noted in Sect. 6.3.1, we obtained this information from Judit Németh, and an analysis of the implications of what we learned formed the basis of the writeup of the talk that Sam Aronson gave about our work at the 1986 Moriond meeting (Talmadge et al. 1986).

An oversight which had the potential to cause problems was an initial lack of awareness of the work of both Renner (1935), Bod et al. (1991) and later Kreuzer (1968). As discussed in Fischbach and Talmadge (1999), Renner was a student of Eötvös who repeated the EPF experiment in 1935. Because he claimed higher sensitivity than the EPF experiment, yet saw no effect, this could have doomed our paper at the outset. Fortunately, we eventually became aware of the careful analysis of the Renner paper by Dicke (1961) and Roll et al. (1964), in preparation for their own experiment (Roll et al. 1964). These authors pointed out various inconsistencies in Renner's results which rendered them unreliable, a conclusion which Renner

---

[8] See footnote on "anti-gravity" in Sect. 6.3.1.

himself confirmed to Dicke (Fischbach and Talmadge 1999, p. 138). A brief note to this effect is contained in Ref. 7 of our original PRL.

Given the potential significance of Renner's results, had they been correct, it was not surprising that we re-engaged with Dicke on this question, in the course of learning more about the locations of the EPF and Renner experiments (see Appendix 3). As can be seen from Dicke's letter of June 27, 1986, he had shown that Renner's errors were too small because Renner failed to account for the fact that his measured values were not independent, since each datum was used more than once. Dicke then goes on to note that although Renner claimed that this procedure was the same as that used by Eötvös, the EPF data seem to be statistically consistent. This agrees with the conclusion we arrived at in our PRL, and in our subsequent more detailed analysis (Fischbach et al. 1988).

The 1968 experiment of Kreuzer (1968), of which we were unaware at the time of our original PRL, was originally conceived as a test of the equality of active and passive gravitational mass. However, it can also be interpreted as a test for an intermediate range force, as was pointed out by Neufeld (1986). Fortunately, the resulting upper limit inferred from the Kreuzer experiment was compatible with the EPF result.

An oversight which was both more significant and more personal was our failure to refer to the seminal papers by Yasunori Fujii (1971, 1972, 1974, 1975, 1981) and Fujii and Nishino (1979). These formed part of the motivation for the geophysical determination of the Newtonian constant G_0 by Stacey and Tuck which in turn motivated our own work. Shortly after our PRL appeared I received a polite note from Fujii pointing out this connection, which I subsequently confirmed in a conversation with Frank Stacey. What Fujii had shown was that in the dilaton theory he was proposing the effective gravitational constant G_0 at laboratory distances could differ by a factor of 4/3 from the constant G_∞ that would describe planetary motion (see Appendix 1). The Fujii papers strongly motivated the work of Stacey and Tuck, which at the time of our PRL was in fact indicating a difference between G_0 and G_∞, and this in turn stimulated our work as we have noted above. Given the clear link between Fujii's work and ours, his paper clearly should have been cited.

Interestingly, in the years prior to our EPF analysis I had compiled a bibliography of relevant interesting papers (Fischbach et al. 1992), and I later found that Fujii's paper in *Nature* (Fujii 1971) was in that bibliography. The same self-imposed time pressure described above ensured that I never consulted this bibliography while drafting our paper, which accounts for our neglect of his paper. I immediately responded to Fujii and apologized. Subsequently I went to some lengths to correct my oversight by detailing the significance of his work in both our review in *Annals of Physics* (Fischbach et al. 1988) and in our book (Fischbach and Talmadge 1999). Eventually we met and became colleagues and friends. We collaborated on a paper (Faller et al. 1989), and during the subsequent years I had the pleasure of being his guest on several visits to Japan.

If it seems surprising that I was upset at missing a single reference in a single paper, my reaction reflects what has always been a firm commitment of mine to fairly credit the work of others, as I would hope they credit my own.

In the category of shortcomings that were not our fault, Ref. 7 of our PRL contains two very unfortunate typographical errors, which were not present in our original manuscript. In order to speed up the publication process, *Physical Review Letters* did not send galley proofs of accepted papers before publication, and hence we had no opportunity to correct these errors. For the record the correct references, as they should have appeared in our paper, are R.H. Dicke: Sci. Am. **205**, 84 (1961) and P.G. Roll, R. Krotkov, and R.H. Dicke: Ann. Phys. (N.Y.) **26**, 442 (1964). The error in the first of these references was particularly embarrassing, especially given the gracious response of Professor Dicke to our paper. Although I apologized to him, he indicated that this was unnecessary since, as one of the referees, he had seen the original manuscript and knew that we had cited him correctly.

Finally, a point which we failed to comment upon, but which arose in subsequent questions, was the role of the brass vials themselves. Specifically, what would the EPF data look like if the samples were taken to be the combination of the brass vials and their contents. Intuitively we had assumed that since the vials were presumably all of the same composition, their contributions would cancel when measuring acceleration differences. Nonetheless this was a question which needed to be addressed in detail, and we did so in our review (Fischbach et al. 1988) by introducing the distinction between "reduced" and "composite" samples, where composite referred to samples when the brass vials were included. As we anticipated, the statistical significance of the EPF results remained unchanged, thus reflecting our original intuition that the contribution from the vials essentially canceled.

6.3.6   Experimental Signals for Hyperphotons

One of the questions that I had been concerned with in the weeks following the submission of our paper to PRL was the possibility of directly detecting the hyperphotons γ_Y, the presumed quanta mediating the field which we had postulated as the source of the EPF result. It had been noted earlier by Weinberg (1964) that branching ratios for decays into hyperphotons can become quite large for reasons discussed below. The EPF results thus motivated us to revisit this question with the aim of relating the hyperphoton coupling constant f in (6.32) (see Appendix 1) implied by the EPF data to existing limits on kaon decays.

Much of the work to be described below was completed before the publication of our PRL on January 6. However, as a consequence of the (previously unexpected) attention following January 6, work on the decays into hyperphotons was interrupted for approximately two weeks. At that point we came to realize that the public attention being devoted to our PRL could stimulate others to raise the same question about constraints implied by decays into hyperphotons. I decided to stay home for part of each day in order to complete the work which Sam Aronson, Hai-Yang Cheng, Wick Haxton, and I had already started. As it turns out our concerns were completely justified: We submitted our paper to *Physical Review Letters* and it was

Fig. 6.8 Decays of $K^\pm$ (**a**) and K_S^0 (**b**) into hyperphotons (Aronson et al. 1986)

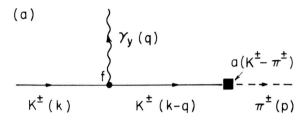

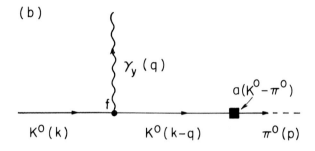

received on January 27. Similar papers, arriving at roughly similar conclusions, were received by *Physical Review Letters* from Suzuki (1986) on January 20, and by *Physics Letters B* from Lusignoli and Pugliese (1986) on January 28, and from Bouchiat and Iliopoulos (1986) on January 29.

Our idea, presented in Aronson et al. (1986), was to examine the decays $K^\pm \to \pi^\pm + \gamma_Y$ and $K_S^0 \to \pi^0 + \gamma_Y$ shown in Fig. 6.8. As seen in the rest frame of the decaying kaons, conservation of linear and angular momentum strictly forbids decays into massless photons, but allows decays into massive hyperphotons. Since the coupling constant f in Fig. 6.8 is small, the probability of a detector actually responding to γ_Y is also small. Hence the signal for $K^\pm \to \pi^\pm + \gamma_Y$ or $K_S^0 \to \pi^0 + \gamma_Y$ would be the appearance of a $\pi^\pm$ or π^0 of energy $m_K/2$, corresponding to $|\boldsymbol{p}_k| = 227\,\text{MeV}$, not accompanied by any other detected particles. The results of a detailed calculation gives the branching ratio (Aronson et al. 1986)

$$\frac{\Gamma(K^\pm \to \pi^\pm + \gamma_Y)}{\Gamma(K^\pm \to \text{all})} = (4.7 \times 10^{14}\ \text{eV}^2)\frac{f^2/e^2}{m_Y^2}\ , \tag{6.25}$$

where e is the electric charge. We see from (6.25) that for $m_Y = 1 \times 10^{-9}\,\text{eV}$, corresponding to $\lambda = 200\,\text{m}$, the branching ratio can be large enough to imply interesting constraints on f^2 or α in (6.1). (The relationship between f^2 and α is discussed in more detail in Appendix 1.) Specifically, using the then-existing limits from Asano et al. (1981, 1982), we found

$$\left|\frac{\alpha}{1+\alpha}\right|\left(\frac{\lambda}{1\,\text{m}}\right)^2 \le 4.7\ . \tag{6.26}$$

A more detailed discussion of decays into hyperphotons can be found in Fischbach and Talmadge (1999), which includes later calculations of the branching ratios $K^{\pm} \rightarrow \pi^{\pm} + \gamma_Y$. Notwithstanding the various theoretical uncertainties that arise in calculating $a(K^{\pm} - \pi^{\pm})$ in Fig. 6.8, the overall conclusion that emerged from the original analysis was that it would have been difficult to simultaneously account for the ABCF data on the energy dependence of the $K^0 - \overline{K}^0$ parameters and the EPF data, while at the same time incorporating the constraints from $K^{\pm} \rightarrow \pi^{\pm} + \gamma_Y$. Of course this assumes that all the claimed effects arise from a single new vector field, and so models with additional new fields are not necessarily excluded.

6.3.7   Visit to Hungary

In the period following publication of our PRL, I received a large number of invitations to speak both in the United States and abroad. Several of these stand out in my mind, particularly my visit to the Eötvös University[9] in Budapest, Hungary May 12–14, 1987. This was arranged by George Marx and included an award to me by the University recognizing my contributions to promoting the importance of the work of Baron Roland von Eötvös. I had several goals in mind, apart from presenting a public lecture on our reanalysis of the EPF experiment and its implications. To begin with, I wanted to determine where in the university EPF had actually carried out their experiment as this would help us to assess the impact of the local mass distribution on the EPF results (see Sect. 6.3.1). Second, I wanted to examine the actual EPF balances which were located in the Geophysical Museum in Tihany, Hungary near Lake Balaton.

As I recall, George and I spent the better part of two days exploring various possible sites. These were signaled by the presence of "Cleopatra's needles", stone piers approximately 1 m on a side, sunk into the ground, presumably to reduce the effects of vibrations. Not surprisingly some of these piers were totally or partially obscured by subsequent construction. Nonetheless we were able to identify likely sites, and this led to an eventual publication (Bod et al. 1991). This reference contains much useful historical material relating to the site of the EPF experiment, as well as additional details on the experiment itself. In the end, we were able to reach a consensus on the likely locations of the EPF experiment, aided by additional input from Jeno Barnothy (see below), Peter Király, Adam Kiss, L. Korecz, A. Körmendi, Judit Németh (see Sect. 6.3.1), and Gábor Palló.

The correspondence in Appendix 3 includes an exchange with Dr. Jeno Barnothy who was a professor at the Eötvös Institute at the University of Budapest from 1935 to 1948, and a colleague of Pekár. Dr. Barnothy, and his wife Dr. Madeleine Barnothy, had retired to Evanston, Illinois the location of Northwestern University,

---

[9]Loránd Eötvös University was founded in 1635, and took the name of its famous one-time teacher in 1950.

and he had contacted me shortly after the publication of our paper. Since Evanston was only a 2.5 h drive from Purdue, I arranged to visit Jeno and Madeleine, and as a result he was able to confirm the locations of the experiments of both EPF and Renner.

Our visit to the Geophysical Museum was even more informative and led to a deeper appreciation of the design of the Eötvös balances. I took a number of pictures and made several drawings of the balances. These led to the diagram shown on p. 133 and the cover of our book (Fischbach and Talmadge 1999). Most notably, the balances contained thermometers which were riveted to the balances, a detail which was not evident in the drawing of the balance contained in Dicke's article in *Scientific American* (Dicke 1961). The significance of the thermometers to us was that Eötvös evidently paid close attention to temperature as a possible systematic influencing their results. From a historical point of view this is of interest in connection with Dicke's proposal that air currents produced by a temperature differential could have accounted for the EPF results. As we have already noted, the Dicke model is not supported by the EPF data, as discussed in Fischbach et al. (1988), and in Sect. 6.2.5. Along with Clive Speake's observations on the significance of the Ag–Fe–SO$_4$ datum, it is clear that EPF did indeed pay close attention to possible systematic influences on their results. In my view this makes their published non-null results even more compelling, and possibly explains why their original results were not published in Eötvös' lifetime.

The trip to Hungary was exciting for an additional reason. Shortly before I left for Budapest I was contacted by *National Geographic* (see Sect. 6.4.3) in connection with the story which John Boslough was working on, and which eventually appeared in the May 1989 issue (Boslough 1989). *National Geographic* is well known for its photography, and they were interested in some photos of me to accompany the story. Given the fact that I was enroute to Budapest it was arranged that a photographer, Adam Woolfitt, would meet up with us in Budapest, which he did. George, Adam, and I drove together to the museum at Tihany. Adam took a large number of photos, and one of them did in fact make it into the story. Adam graciously sent me some of the others, which were quite useful to Carrick and me in writing our book (Fischbach and Talmadge 1999).

6.3.8 The Air Force Geophysics Laboratory Tower Experiments

At the time our PRL appeared the United States Air Force maintained two laboratories dedicated to geophysical research, one located at Hanscom AFB in Bedford, Massachusetts and the other at Kirtland AFB in New Mexico. (At present there is a single site at Kirtland.) The Hanscom site was then headed by Don Eckhardt who, along with Andrew Lazarewicz, Anestis Romaides, and Roger Sands, organized an experiment to measure the local acceleration of gravity g up a tall tower. In some sense this was the mirror image of the original experiment of Stacey and Tuck, and was in principle sensitive to deviations from the inverse-square law over the same 1 km range. Based on conversations I have had with Don, it seems that the initial motivation for these experiments was to improve the upward continuation of gravity measurements taken at the surface to altitudes where they would be relevant for missile inertial guidance systems. In fact Don had been planning a balloon experiment to measure gravity at altitudes up to $\sim$100,000 ft. It is not hard to imagine that then-existing inertial guidance systems might be sensitive to deviations from Newtonian gravity at a level suggested by the data of Stacey and Tuck, and/or our EPF analysis. (However, rumors at the time that Air Force missiles were missing their targets in test firings by more than had been expected, have not been confirmed to me by Don.)

In any case, the conceptual framework was clear: By measuring Newtonian gravity over a sufficiently large area surrounding a tall tower, one could use Newtonian gravity to extrapolate these data and predict what g should be going up the tower. These predictions would then be compared to the actual measurements on the tower carried out by a sensitive Lacoste-Romberg gravimeter which Anestis and Roger carried up the tower. Any discrepancies between these measurements and predictions could then be a signal for deviations from the inverse-square law.

Eckhardt and his collaborators at the Air Force Geophysical Laboratory (AFGL) carried out their first experiment using the 600 m WTVD television tower in Garner, North Carolina, and initially found what they characterized as a "significant departure" from the predictions of Newton's inverse-square law. Their quoted departure, "approaching $(500 \pm 35) \times 10^{-8}$ m/s^2 at the top of the tower," was published in *Physical Review Letters* on June 20, 1988, a few weeks before the Fifth Marcel Grossmann meeting in Perth Australia (Eckhardt et al. 1988) (see also Sect. 6.4.3). Since the sign of their effect corresponded to a new "attractive" force, in contrast to the repulsive fifth force implied by the EPF data, Eckhardt and collaborators characterized their result as the discovery of a new "sixth force", and this was one of the exciting stories at the Marcel Grossmann meeting.

However, the results of the tower experiment, along with those of the original Stacey experiments, were soon called into question by Bartlett and Tew (BT) (1989a, 1989b, 1990). In brief, BT noted that the evidence for non-Newtonian gravity reported in each case could have arisen from "terrain bias", wherein the gravity measurements in the vicinity of each site did not accurately reflect the actual terrain at the site. The AFGL collaboration refined their analysis, and eventually withdrew their claim of evidence for non-Newtonian gravity (Jekeli et al. 1990).

In 1990 the AFGL (renamed the Phillips Laboratory) began another tower experiment, this time at the 610 m WABG tower in Inverness, Mississippi. By this time Carrick was being supported as a postdoc by AFGL/Phillips, thanks to the efforts of Don Eckhardt, so he and I were invited to join this new effort.

GPS location required that at least four satellites be in view, but at the time of our experiment in the early 1990s this was not always the case. However, Anestis had a program which told us when at least four satellites could be seen, and this sometimes required us being up late at night or getting up early in the morning. To avail ourselves of GPS, a circular grid was defined by Anestis and Roger extending out to approximately 10 km from the WAGB tower.

One of the tasks assigned to Carrick and me during our first visit in November 1991 was to install platforms at the 128 designated sites at which ground-level gravity measurements were made. Given the "terrain bias" effects that had been problematic at the previous WTVD site, we were absolutely committed to installing these platforms exactly where they were supposed to be as specified on a map, irrespective of how unwelcoming these sites might be for one reason or another. Some of these were in wetland areas, and others were near catfish ponds whose owners were not always thrilled at having strangers on their property. Since Mississippi has a strong military tradition, our encounters with local residents on whose property we were carrying out our work were generally pleasant, once they learned we were on an Air Force project. In our subsequent visits in December 1991 and the Spring of 1992, when GPS and gravity measurements were actually performed at these sites, we were faced with the problem of carrying relatively expensive equipment to these sites, hoping that we would not drop any of this equipment into some body of water.

However, things did not always go well. The WABG tower was located in the Mississippi delta region, whose soil formed a fine wet clay that locals called "gumbo". On more than one occasion our military "humvee" got stuck in the "gumbo", as did one of our rental vehicles. On another occasion a prison work gang ran over one of our sites, located in plain view in front of a church, and destroyed the car battery running the GPS equipment.

In addition to problems with the ground survey, we also experienced problems with the tower gravity measurements to which the ground measurements were to be compared. Given the extreme sensitivity of the Lacoste–Romberg gravimeters that we were using, vibrations of the tower due to wind precluded obtaining useful measurements unless the wind speeds were very low, typically less than 5 km/h. Although this meant that days went by when gravity measurements up the tower could not be made, eventually measurements were made at the lower levels on days that were sufficiently calm. Additionally we experienced radio-frequency interference with our measurements, which was not surprising given that we were on a television tower. This problem was eventually resolved by moving our equipment to a slightly different position on the tower. Finally there was always the problem of lightning strikes while somebody was on the tower. These were potentially problematic given the very slow speed of the elevator used to move up or down the tower. The group managed to see storms moving in our direction in sufficient time to get down, and nobody was hurt.

In the end, the choice of the WABG tower was a good one. The flatness of the terrain, combined with the stability of the WABG tower (and the cooperation of its owners), allowed us to significantly improve on the earlier results from the WTVD tower. Our results led to agreement with Newtonian gravity, represented by the largest difference being

$$(\text{observed} - \text{discrepancy}) = (32 \pm 32)\ \mu\text{Gal} \text{ @ } 56\,\text{m}, \qquad (6.27)$$

where $1\,\mu\text{Gal} = 10^{-8}\,\text{m/s}^2$ (Romaides et al. 1994, 1997).

The null result from the WABG tower experiment is supported by two other tower experiments, which were carried out at approximately the same time: Speake et al. using the 300 m NOAA meteorological tower in Erie, Colorado (Speake et al. 1990), and Thomas et al. using the 465 m tower at Jackass Flats, Nevada (Thomas et al. 1989; Kammeraad et al. 1990). Although all three tower experiments arrived at a null result with respect to possible deviations from Newtonian gravity, they demonstrated for the first time that such experiments could in fact be carried out with sufficient sensitivity to provide useful α–λ constraints over the 1 km distance scale (Fig. 6.9), as was first suggested by Don Eckhardt. Further discussion of these experiments can be found in Fischbach and Talmadge (1999).

Examination of Fig. 6.9 reveals an interesting fact that I incorporated into all of my early fifth force talks. As indicated in the figure caption, the only values of α and λ that were allowed by the existing data in 1981 or 1991 are those falling below the corresponding shaded regions in the figure. We then see that as late as 1981 (almost 300 years after Newton), α could be as large as 0.1 (corresponding to a 10 % discrepancy with Newtonian gravity) over a distance scale of approximately 10 m, and still be consistent with experiment. I called this at the time the "10–10" mnemonic (10 % at 10 m), and it came as a big surprise to my audiences, particularly since 10 m seems to be a distance scale that is readily accessible to laboratory

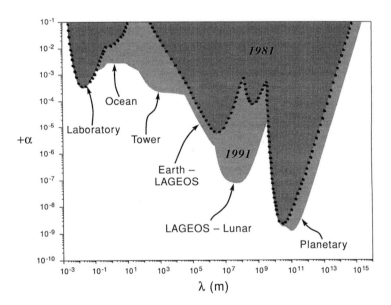

Fig. 6.9 Long-range constraints on α as a function of λ as of 1991 (Fischbach and Talmadge 1992b). For each of the two regions labeled 1981 and 1991 and above them, the *shading* denotes values of α and λ which are excluded by the indicated experiments or analyses. The *dotted curve* denotes the envelope of allowed values as of 1981 and, as indicated in the text, the 1981 data allowed for a discrepancy with Newtonian gravity of 10 % for distances scales of order 10 m. See also Talmadge et al. (1988)

experiments. This in turn relates to another question, which is directly related to the tower experiments: why can't we carry out a precise measurement at one particular distance scale and have it apply to all scales?

To answer this question consider a possible fifth force contribution to the precession of the perihelion of an elliptical orbit about the Sun of planet P with semi-major axis a_P. It is straightforward to express the precession angle $\delta\phi_a$ in terms of α, λ, and a_P (Fischbach and Talmadge 1999, p. 114):

$$\delta\phi_a \approx \pi\alpha \left(\frac{a_P}{\lambda}\right)^2 e^{-a_P/\lambda} . \tag{6.28}$$

One can show that for a given value of α, $\delta\phi_a$ reaches a maximum when $a_P/\lambda = 2$, and vanishes when either $a_P/\lambda \to \infty$ or $a_P/\lambda \to 0$. In the former case the range of the fifth force is too short for the Sun to influence the planet. In the latter case the range of the fifth force is so long, that an observer at a_P would experience a predominantly $1/r^2$ force, which causes no precession of the perihelion. Similar arguments apply to other inverse-square tests of Newtonian gravity, such as the limits labeled "Laboratory" in Fig. 6.9, which are from Spero et al. (1980) and Hoskins et al. (1985). The preceding discussion explains why the most sensitive limits on α at a given λ are obtained when the size of the system being studied (the analog of a_P) is close to the magnitude of λ being studied. This also helps to explain why the constraints arising from planetary data in Fig. 6.9 are so much more restrictive than those at other scales: there are simply many more data available at solar system length scales than elsewhere. For further discussion, see Talmadge et al. (1988).

It should be noted that the situation regarding composition-dependent fifth force searches is quite different since a very long-range (i.e., $1/r^2$) force which was composition-dependent would still show up as a deviation from the predictions of Newtonian gravity. This is, of course, precisely the theory that Lee and Yang were testing in their classic paper (Lee and Yang 1955). The fact that a composition-dependent deviation from Newtonian gravity can be present and detected, even for a long-range ($1/r^2$) force, explains why the resulting constraint curves look qualitatively different from the α–λ curves describing composition-independent searches.

6.3.9 An Electromagnetic Fifth Force?

Just as a fifth force coupling to baryon number could produce deviations from the predictions of Newtonian gravity, so one could imagine another type of fifth force coupling to electric charge, whose presence could be detected via deviations from the predictions of Maxwell's equations or quantum electrodynamics. This possibility was raised by Bartlett and Lögl (BL) (1988) who considered the implications of a potential $V(r)$ between two electric charges e having the form [in analogy to the gravitational fifth force potential energy (6.1)]

$$V(r) = \frac{e^2}{r} \left(1 + \beta e^{-r/\lambda}\right) , \tag{6.29}$$

where β is the dimensionless strength relative to the Coulomb force, and λ is the range. Although it is natural to assume that electromagnetism has been sufficiently well tested over all distance scales as to allow only very small values of β, BL pointed out that there was in fact a region $\lambda \approx 1\,\mu m$, where limits on β were relatively poor. As in the case of gravity, this "gap" arises because there are fewer systems of this size which are readily accessible to experiments.

The paper by Bartlett and Lögl led to a series of papers by our group Krause et al. (1994), Fischbach et al. (1994), and Kloor et al. (1994), which was part of my student Harry Kloor's physics Ph.D. thesis (Sect. 6.7). In the process of deriving new geomagnetic limits on the photon mass, using data on the Earth's magnetic field supplied by Bob Langel (who was then on sabbatical at Purdue) (Fischbach et al. 1994), Harry became interested in other limits which appeared to be more restrictive. He eventually found that the then-existing best limit quoted by the Particle Data Group (PDG) could not be justified. We subsequently informed the PDG, and this eventually led to me becoming for a time the consultant to the PDG for the photon mass (PDG 1998).

6.4 Reflections

6.4.1 The Moriond Conferences

No organizational effort contributed more to searches for non-Newtonian gravity (and other related exotic phenomena) than the Rencontres de Moriond under the leadership of J. Trân Thanh Vân. Following the publication of our original paper in January 1986, Sam Aronson was invited to give a talk on our work at the Twenty-First Rencontre de Moriond, which took place from March 9–16, 1986. The meeting was held at Les Arcs, which is a ski resort conveniently located approximately 3 h by bus from Geneva and CERN, where Sam was working at that time, on leave from Brookhaven. Sam gave a general presentation of our work, and the write-up which appeared in the Proceedings (Talmadge et al. 1986) focused on the issue of local mass anomalies, which we have discussed above.

The Moriond organization had a workshop scheduled for January 24–31, 1987 entitled "New and Exotic Phenomena", which would also take place at Les Arcs. In addition to sessions on such (then) exotic topics as CP-violation, dark matter, neutrino mass and oscillations, they had decided to include a session devoted to the fifth force. By that time there were already a number of experiments underway, and representatives of some of these efforts were present. These included Frank Stacey, Fred Raab (Eöt-Wash experiment), Peter Thieberger, Pier Giorgio Bizzeti, Riley

Newman, and Kazuaki Kuroda. Additionally there were related talks by Mike Nieto and Bill Fairbank, on tests of the gravitational acceleration of antimatter, and theory talks by John Moffat, Bob Holdom, and Alvaro de Rújula.

The schedule of the Les Arcs meeting, and other Moriond meetings, was typically as follows: talks began at 8:00 and lasted to 12:00. There followed a break until 16:00 which included lunch and time for skiing. Talks then resumed until 20:00, followed by dinner. Since skiing is not one of my better sports, I welcomed the opportunity to improve my skills, with the aid of some instruction arranged by the Moriond organization. The break from 12:00 to 16:00 encouraged informal physics conversations both on and off the slopes.

Dinners provided an opportunity for more detailed discussions among participants with common interests. One evening, while several of us from the fifth force session were having dinner together, the conversation drifted to criticisms of the Eöt-Wash experiment as described by Fred Raab, led by Frank Stacey. Fred was reporting a null result whereas Frank's anomalous result was part of the motivation for our original PRL paper. Fred stuck to his guns despite intense questioning by Frank and others, myself included. In the end it turned out that Fred was correct, whereas Frank withdrew his published anomalous results, as noted in Sect. 6.3.8.

Towards the end of that week there was an organizational meeting called to plan for the next Moriond Workshop in January 1988. I was invited to that meeting which I interpreted as a sign that the quality of the fifth force talks had met with general approval from the group. This view was not unanimous, with Felix Boehm expressing some concern that this work was still highly speculative. Nonetheless the decision was made to go ahead with a larger fifth force session in 1988: the workshop title was to be "5th Force Neutrino Physics".

Measured in terms of the experimental effort devoted to fifth force experiments, the 1988 workshop was the high-water mark, and gave this nascent field a major boost. As the organizer primarily responsible for arranging the fifth force session, I worked hard to cover as many of the ongoing experiments, or proposed experiments, as possible. In the end there were 26 talks in the fifth force session, which I opened with an overall introduction to current research. The written version of my talk, which appeared in the Proceedings (Fischbach and Talmadge 1988), contained an additional feature which we included in subsequent talks: this was a list of all the experiments known to us as of April 1, 1988, broken down by category. The 1988 tabulation listed 45 experiments, which was quite remarkable considering that only two years had passed since the publication of our original paper in PRL.

Support by Rencontres de Moriond for research related to the fifth force continued in subsequent years. The 1989 January workshop also included a session on the fifth force with 16 talks, and the 1990 January workshop featured 13 talks which were fifth force related. By 1993 it had become clear that virtually all modern experiments were finding null results, the lone exception being Peter Thieberger's floating ball experiment (Thieberger 1987). The January 1993 Workshop included a session on gravitation, with 11 talks on tests of the Equivalence Principle, the rubric which to some extent has superseded the fifth force in searches for composition-dependent deviations from Newtonian gravity.

In 1996, the tenth anniversary of the publication of our original paper in PRL, I was invited to give one of two "special lectures", which are meant to be somewhat broader in scope so as to be understandable to all of the participants at the workshop. I chose as the title of my talk "Ten Years of the Fifth Force", and in that talk Carrick and I reviewed what we had learned in the previous 10 years:

> One can summarize the current experimental situation as follows: There is at present no compelling experimental evidence for any deviation from the predictions of New-tonian gravity in either composition-independent or composition-dependent experiments. Although there are some anomalous results which remain to be understood, most notably in the original Eötvös experiment, the preponderance of the existing experimental data is incompatible with the presence of any new intermediate-range or long-range forces.

Notwithstanding that somewhat disappointing conclusion, there was much that had been learned in the preceding decade. To start with many novel and clever experiments had been carried out and refined during that period.[10] Additionally, a phenomenological framework had been established which characterized most experiments in terms of the parameters α and λ (or ξ and λ) as summarized in Appendix 1. The constraints on α or ξ as a function of λ implied by different experiments could thus be combined on a single common plot as shown in Figs. 1 and 2 in my 1996 Moriond talk (Fischbach and Talmadge 1996). Examination of these plots showed that even by 1996 significant regions of the $\alpha-\lambda$ and or $\xi-\lambda$ planes had been excluded by various experiments, and this trend has continued to the present. The $\alpha-\lambda$ and $\xi-\lambda$ plots have by now become useful tools for theorists in constraining possible new scenarios for physics beyond the standard model. For example, theories involving extra spatial dimensions typically predict deviations from Newtonian gravity over short distances. As discussed in Sect. 6.6, the number of extra dimensions allowed can be constrained by appropriate inverse square law tests carried out over small separations, whose results can be expressed in $\alpha-\lambda$ plots. It is gratifying that such constraints have been included in the Particle Data Group reviews (PDG 2014).

My guess is that searches for deviations from Newtonian gravity would have had a much more difficult time becoming part of mainstream physics, had it not been for the Rencontres de Moriond and the credibility they lent to such efforts. In addition to the meetings themselves, and the opportunities they provided for interactions among the participants, the Proceedings from each meeting played an important role by collecting together many of the early experimental results and theoretical ideas. In the early years these were usually edited by Orrin Fackler and Vân himself. We all owe this group, under the leadership of J. Trân Thanh Vân, and more recently Jacques Dumarchez, a deep sense of gratitude.

In March of 2015, on the occasion of the 50th anniversary of the Rencontres de Moriond, I was asked by Jacques Dumarchez to give another general interest

---

[10]We have learned a great deal from these experiments, for example, that great care must be exercised in continuing gravity measurements taken at the surface of the Earth upward to towers or downward to mines.

"special lecture" to a joint session of the two workshops that were meeting at the same time. I chose as the title of this talk, "Rencontres de Moriond and the 5th Force". Aided by my long-time collaborator Dennis Krause, we assembled a review of the entire history of the fifth force as recorded by the proceedings of the Rencontres de Moriond over the years since 1986. Following my talk, Jacques made the interesting observation that not only had Moriond given a boost to the fifth force but, reciprocally, the fifth force had helped Moriond by motivating the Rencontres to expand into new areas beyond particle physics. These included gravitation and atomic physics, which have become increasingly exciting areas, but which had not been regular topics prior to 1986.

6.4.2 Some Amusing Moments

The *New York Times* story, and the associated depiction of the falling coin and feather, spawned a number of amusing moments, some intentional and others not. In the former class was a cartoon published in the *Seattle Post-Intelligencer* shortly after the *Times* story drawn by Steve Greenberg (see Fig. 6.10). This was clearly based on the depiction of the fifth force in the *Times* drawn as opposing gravity, and hence acting as a new "anti-gravity" force (see Sect. 6.3.1). Idella Marx, whose husband Louis Marx had been on the cover of *Time* magazine, had much more experience with the media than I did, and so took it upon herself to obtain the

Fig. 6.10 Cartoon by Steve Greenberg published in the *Seattle Post-Intelligencer* on the fifth force

original of that cartoon for me. I learned from her that the authors of cartoons often sell the originals as an additional source of income. She asked Greenberg to donate the original to me, which he graciously did along with his autograph, and it now hangs in my study.

In the category of unintentional amusing moments spawned by the fifth force is another "coin and feather" story, and its consequences. In 2000 I was nominated by my department head to interview for an assignment with the Thinkwell company of Austin, Texas. This involved filming a series of laboratory demonstrations to accompany an online undergraduate text that they were developing. For each candidate the "interview" consisted of filming a demonstration of the applicant's choosing in which he/she explained the physics behind the demonstration. Naturally I chose the "coin and feather" demonstration, which began with me demonstrating that with air present in the glass tube apparatus, the coin fell faster, as we expected. I then rotated the stopcock on the glass tube and started the vacuum pump to remove the air. Finally I turned the glass tube upside down to demonstrate that in a vacuum the coin and feather fell at the same rate. Except that they didn't! At that instant I responded by blurting out "...because this demonstration *didn't* work this proves that it is a genuine physics demo!" What had happened was that the glass tube had a somewhat unusual stopcock which required another 1/4 revolution to connect to the vacuum pump. I quickly repeated the demonstration and explanation which now worked. Since I was pressed for time, I decided not to edit the film and sent it as is to Thinkwell. To my surprise I was hired for the assignment and, as it turns out, my humorous response to the original failure turned out to have been a net plus in my interview.

As time went on, and it became clear that a fifth force with the characteristics we assumed did not exist, I became known in the family as "...the discoverer of the non-existent fifth force." Naturally, I took this in good spirits, particularly since it fostered a collective sense of humor in our family which we all appreciated. An incident which (almost) happened occurred on October 23, 1993 when my son Jono took the SAT college entrance exam, while many other students across the country took the alternative ACT test. The latter included a reading comprehension section on the fifth force taken from a piece written by Michael Lemonick entitled "Working Against Gravity".[11] I received a number of calls that day from friends and former grad students whose children took the ACT and recognized my name. We can only speculate what Jono's reaction would have been had he taken the ACT rather than the SAT. Would he have answered the question correctly? Would the surprise of being confronted with that question have distracted him and impacted his overall performance? Fortunately we will never know—he did quite well on the SAT and was accepted to Princeton.

An amusing incident which did happen took place during the summer of 1987. We had arranged to meet Jerry and Sharon Lloyd along with their children Brendan and Heidi whom we had met during my sabbatical at UW. Brendan and our son

---

[11] Although the ACT declined my request for a copy of the question, they indicated that the same passage was administered to approximately 308,000 test takers between 1990 and 1999.

Jeremy had become close friends, and so we decided to meet in Durango, Colorado for a week together. One day we decided to take the famous train ride from Durango to Silverton, and we sat in our open gondola car as the collection of five children from the two families scampered from side to side to better view the spectacular scenery. A very staid passenger looked upon the scene with silent—but obvious—disapproval. On the return trip from Silverton to Durango I ended up chatting with him and learned that he was a high school physics teacher from Quebec. One thing led to another, and when he eventually learned that I was a physics professor at Purdue he asked whether I knew the individual who was working on the fifth force. When I acknowledged that I did, he kept asking questions, not quite realizing who I was. Gradually, like those old Polaroid pictures which slowly came into focus, he realized that I was the individual he was asking about. At that point we broke the ice, and we both enjoyed a big laugh.

No discussion of the humor associated with the fifth force could be complete without reference to the spoof written by Lawrence Krauss, which was actually submitted for publication to *Physical Review Letters* shortly following the appearance of our paper. Krauss, who was then at Yale and is now at Arizona State University, distributed a preprint (which I received) entitled "On Evidence for a Third Force in the Two New Sciences: A Reanalysis of Experiments by Galilei and Salviati." This "paper" is quite funny, but at the time I had no idea that this was actually submitted for publication in PRL. George Basbas, who was the PRL editor at the time, obviously realized this was a spoof, and returned six reports on it, "one [report] for each force." Although the Krauss paper was not accepted by PRL, it was published in 2008 in *Physics Today* (Krauss 2008), along with the six "reports", and is well worth reading.

6.4.3 Fifth Force Stories: Journals vs. Magazines

The publication of the *New York Times* story about our work on Wednesday, January 8, 1986 was preceded by short items on Tuesday evening on NBC radio, and on CBS TV evening news with Bob Schieffer. Following the full story in the *New York Times*, stories also appeared in newspapers all over the world. Given the overwhelming world-wide impact of the *New York Times* story, there can be little doubt that—at least in those days—the *New York Times* exerted an enormous influence in determining which stories were newsworthy. I recall somebody with expertise in such matters opining that virtually every major newspaper in the world must have mentioned this story in the subsequent weeks, including one of my favorites, a newspaper in Iceland. Subsequently, "fifth force" made it into an Icelandic–English dictionary that was being compiled by my friend Christopher Sanders and others. (For the record the translation is "ofurhlðslu kraft/ur" (Hólmansson et al. 1989).)

For many of these stories the journalists/science writers contacted me directly, and I could tell immediately that some were much more eager than others to spend the time to understand the details of what we had done, and what the implications would have been if there really were a fifth force. Among the many newspapers that ran stories in the subsequent weeks and months, I was particularly impressed with both the *New York Times* and the *Los Angeles Times*.

One of the persistent problems in dealing with the popular press was ensuring that they appropriately credited my co-authors: my students Daniel Sudarsky, Aaron Szafer, and Carrick Talmadge, as well as Sam Aronson whose early collaboration with me was the motivation for the EPF analysis. My co-authors on the original (and subsequent) papers were exceptionally talented as individuals and as a group, and their contributions to this paper were as significant as my own. I was particularly interested in seeing to it that Carrick be recognized since this work became the central part of his Ph.D. thesis. Although I had little influence over major publications such as the *New York Times* and the *Los Angeles Times*, they mentioned all of the co-authors of our paper, for which I was deeply grateful. For our local newspaper, the *Lafayette Journal and Courier*, I felt that I could exert greater influence, and I did whenever possible. I recall receiving a call at home on the eve of Rosh Hashanah just as I was leaving for services at our synagogue. I explained to the reporter why I couldn't talk, but he was eager for an interview anyway. So I agreed to meet with him after services at the newspaper, in exchange for a commitment on his part to feature my students in his story. So following services I drove to the newspaper and rang a bell at the particular entrance where we had agreed to meet. By now it was nighttime, and we stood huddled at the entrance to the paper talking in the dark, in a scene that evoked in me images of "Deep Throat" speaking to Carl Bernstein and Bob Woodward.

In dealing with the popular press, whether in the form of newspapers or magazines, I often felt the tension between me as a scientist trained to appropriately cite other researchers whose work motivated my own, and story writers who almost always labor under stringent word limits for their stories. This became more of a problem as other researchers entered the field and made significant contributions of their own, which deserved to be recognized in print.

For me, the most dramatic example of this tension presented itself in the story by John Boslough in the May 1989 issue of *National Geographic* (Boslough 1989). This story was based in part on a dinner in Perth, Australia to which John had invited Eric Adelberger and me. Eric and I were attending the Fifth Marcel Grossmann meeting in Perth, August 8–13, 1988, and John was interested in learning more about both the underlying theoretical ideas (from me), and the experimental situation (from Eric). Following my review of the motivation for our reanalysis of the EPF experiment, Eric gave a nice description of his experiment, emphasizing (appropriately) the many improvements his Eöt-Wash collaboration had made over both the previous RKD and BP experiments. In fact Eric's presentation was so compelling, that I entertained the humorous thought that perhaps the reason why he was not reproducing the EPF results was that his experiment was *too* perfect!

Eventually John's story was completed, and led to my first encounter with "fact checkers", members of the *National Geographic* staff whose job it was to literally check and verify every fact and statement in the story. I was sent a pre-publication copy of the story and asked to verify a number of items directly related to parts of the story relating to me. There were indeed a few minor mis-statements which I pointed out, but my task did not end there: I was asked to replace the existing text with a corrected version that would not take up additional space. In most journals, such a request made by a referee would not be a problem, since space is not usually an issue. However, the changes that were required were mostly ones which would have benefitted from greater elaboration, and hence more space—which I was not allowed. Nonetheless, I worked closely with the two fact checkers to arrive at a compromise, and they were appreciative for my efforts on their behalf.

By this time I had developed a close relationship with the fact checkers over the course of several phone conversations, and so I decided to press them to correct what I felt was an unfortunate omission in John Boslough's otherwise superb story: there had been no mention of the elegant experiment by Peter Thieberger from Brookhaven National Laboratory, which was the very first experimental test of the fifth force idea, and which had in fact found a positive result which could have been interpreted as supporting our EPF analysis (Thieberger 1987). Although subsequent experiments have not found evidence supporting the idea of a fifth force, it is not clear what—if anything—was wrong with Thiebeger's experiment. It became clear immediately that there were two problems that I was facing in trying to include mention of Thieberger's work: Although *National Geographic* and its authors were presumably happy to have me correct aspects of the story as written, they were not inclined to allow me to modify the story by including new material. Additionally, whatever new material I wanted to add would have to come up against the stringent space requirements discussed above.[12] I decided to tackle the second problem first, by compressing a description of Thieberger's experiment down to 26 words. I then found a comparable savings elsewhere in the story, so my suggestion was "word neutral". Although I do not know exactly what happened thereafter, I presume that the fact checkers must have contacted John Boslough and received his approval, which thus solved the first problem. In the end my proposed text appeared in the final published version on p. 570, much to my delight.

6.4.4 John Maddox and Nature

The publication of our PRL occurred during the period when John Maddox was the editor of *Nature*. Although our original paper was not published in *Nature*, John took a keen interest in our work, and wrote several favorable editorials on the subject (Maddox 1986a,b, 1987, 1988a,b,c, 1991). Additionally, he invited Carrick and me to write a review of the field to be published in 1991, which would have allowed us to adopt the mellifluous title "Five Years of the Fifth Force". Unfortunately, various delays ensued, so that by the time the review appeared in 1992 (Fischbach and

---

[12]Recall that, in contrast, PRL allowed us to exceed their nominal length allowance in order to address a question raised by Dicke, as we note in Sect. 6.2.5.

Talmadge 1992a), we were forced to change "Five" to "Six". Based on conversations I had at the time it is clear that the prestige of *Nature* was such that our review, along with John's editorials, gave our work and the field in general a significant boost at a critical time.

6.5   My 1985–1986 Sabbatical at the University of Washington

As noted above I had been invited to spend the 1985–1986 academic year at the Institute for Nuclear Theory (INT) at UW, mostly due to the efforts of Wick Haxton who had been an Assistant Professor at Purdue before joining the UW faculty. I was warmly welcomed by the INT faculty, including Ernest Henley, Larry Wilets, and Jerry Miller among others. The INT faculty went well beyond what would have ordinarily been expected of them. For example, INT agreed to pay to have the snakewood sample chemically analyzed, and to pay the INT secretary JoAnn LaRock overtime to come in on a weekend to help me answer the dozens of letters I received following the publication of our paper. More importantly, the UW faculty viewed our paper seriously to the extent that several faculty undertook experiments to test the implications of our PRL paper. Most notably, Eric Adelberger, a well-known and highly respected nuclear physics experimentalist, and now member of the National Academy of Sciences, established the "Eöt-Wash" collaboration (a pun on the Hungarian pronunciation of "Eötvös"). He has by now become the world's leading experimentalist in searching for deviations from the predictions of Newtonian gravity. Eric was joined over the years by Jens Gundlach, Blayne Heckel, Fred Raab, and Chris Stubbs among others, and the work of this group continues to date. In addition, Paul Boynton entered the field, and over time joined forces with Riley Newman and Sam Aronson. The work of this group also continues to date. Among the efforts at the time, Dick Davisson (a son of Nobel Laureate Clinton Davisson) designed an extremely clever test for a composition-dependent fifth force using a MACOR sphere suspended in water by means of an "inverse Cartesian diver". Unfortunately this experiment was never completed.

During my stay at the UW I enjoyed the many conversations I had with Eric Adelberger and other members of the Eöt-Wash collaboration as well as with Paul Boynton and his group, and with Dick Davisson. Although I was never an actual participant in any of the UW experiments, I kept in reasonably close contact with the various experimental efforts. So it came as no surprise to me when I received a call one evening from Paul Boynton urging me to return to the Physics Department, because he was seeing evidence for a fifth force. I was almost ready to leave home, a 45 min drive to UW, when I sensed that he was just trying to test me, so that in the end his call was just an attempted prank.

However, some time later the Boynton group did in fact claim to see evidence for a fifth force (since withdrawn), and their paper was accepted for publication in

PRL (Boynton et al. 1987). Having learned from Paul that the American Institute of Physics (AIP) was preparing a press release on this experiment, I quickly prepared my own spoof press release by modifying AIP letters I had received, along with a covering letter (Appendix 6). I arranged to have it sent to Paul from New York, the home of AIP, so that it would look authentic. Having been myself the subject of a number of stories in the press, I began with the usual stiff formal language, but then gradually introduced a "humor gradient", where each successive sentence was increasingly implausible. Paul was apparently taken in until the very end, and was on the verge of contacting the AIP and complaining when he realized that this was a spoof, and we all had a big laugh.

6.6 Short-Distance Searches for a Fifth Force

Just as our book (Fischbach and Talmadge 1999) was being completed, a new set of ideas was emerging leading to the prediction of new macroscopic forces manifesting themselves over very short distances (Antoniadis et al. 1998; Arkani-Hamed et al. 1999; Randall and Sundrum 1999). Broadly speaking these forces are a reflection of the hypothesis that we live in a world with n-additional compact spatial dimensions, which could manifest themselves over scales from submillimeter to angstrom distances, or even smaller. As can be seen from Figs. 6.11 and 6.12, the limits on the strength α as a function of the range λ of a new force become increasingly less stringent as λ gets smaller. As a result current limits on new forces at the sub-micron level allow for the existence of new macroscopic forces significantly stronger than gravity.

If we denote the scale of the n additional spatial dimensions as r_n, then in typical theories the effective gravitational potential $V(r)$ between two point sources is given

Fig. 6.11 Limits on the fifth force strength $|\alpha|$ for $\lambda \geq 1$ cm from laboratory, geophysical, and astronomical measurements (Adelberger et al. 2009)

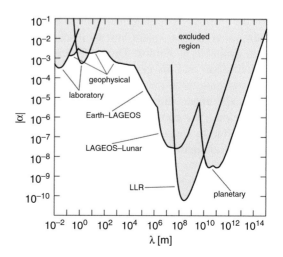

Fig. 6.12 Limits on the fifth
force strength $|\alpha|$ for
$\lambda \leq 0.1$ mm from
short-distance force
experiments along with
predicted strengths from
various theories (Chen et al.
2014). "IUPUI" labels
constraints coming from
experiments with Ricardo
Decca and Daniel López
utilizing "iso-electronic"
effect experiments

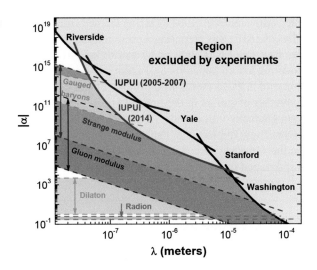

by (Floratos and Leontaris 1999; Kehagias and Sfetsos 2000)

$$
V(r) = \begin{cases} -\dfrac{G_{\infty}m_i m_j}{r}\left(1 + \alpha_n e^{-r/\lambda}\right) , & r \gg r_n , \\[2mm] -\dfrac{G_{4+n}m_i m_j}{r^{n+1}} , & r \ll r_n . \end{cases} \tag{6.30}
$$

Here α_n is a dimensionless constant, which would depend on the number of additional spatial dimensions and their compactification, $\lambda \sim r_n$, and G_{4+n} is the more fundamental Newtonian constant in the $(4 + n)$-dimensional space-time. In a theory where all the additional dimensions are of the same size, and have toroidal compactification, then $\alpha_n = 2n$.

It follows from the preceding discussion that the signal for new physics implied by the presence of additional spatial dimensions would be a violation of the Newtonian inverse-square law, as discussed in Appendix 1. Given the facilities then available at Purdue, Dennis Krause and I joined with my colleague Ron Reifenberger and his graduate student Steve Howell to carry out an experiment using atomic force microscopy (AFM) at the nanoscale (Fischbach et al. 2001). This experiment, and all subsequent experiments that we carried out at this scale (see below), was complicated by the Casimir force, the attractive force between two bodies due to vacuum fluctuations (Bordag et al. 2015; Simpson and Leonhardt 2015). Although this force is negligible for macroscopic experiments, it is the dominant known force between electrically-neutral non-magnetic bodies at the sub-micron separations which were of interest to us, and we eventually chose to deal with it in two complementary ways. The conceptually simplest was to calculate the Casimir force, and subtract it from the experimentally measured force. This was the heroic task carried out by our colleagues Vladimir Mostepanenko and Galina Klimchitskaya (Decca et al. 2005a). The second approach utilized what we called the "iso-electronic" effect in which one searches for force differences between dissimilar materials with similar electronic properties (and hence the same Casimir force), for example, two isotopes of the same element (Krause and Fischbach 2002; Fischbach et al. 2003). When we were eventually joined by our colleagues Ricardo Decca and Daniel López, a better technique emerged: simply measure the force difference between a probe and *any* two dissimilar samples coated with a common ~150-nm thick layer of gold (Decca et al. 2005b). Since the Casimir force is primarily a *surface* effect, this layer is sufficiently thick to make the Casimir force between the samples and probe the same, but thin enough to permit force differences due to new gravity-like interactions which are *bulk* effects.

The experimental and theoretical collaboration among Dennis Krause, Ricardo Decca, Daniel López, Vladimir Mostepanenko, Galina Klimchitskaya, and me has now led to a long series of papers, resulting in the most stringent limits on a Yukawa-type fifth force in the 40–8000 nm range (Fig. 6.12) (Chen et al. 2014). Not surprisingly, these limits still allow new forces many times stronger than Newtonian gravity over short distances, and hence the community is not yet near the point of excluding new forces weaker than gravity, over these distances.

6.7   Our Book: "The Search for Non-Newtonian Gravity"

On April 18, 1986 I received a letter from Robert Ubell at the American Institute of Physics (AIP), copied to Rita Lerner, discussing the possibility of writing a book on the fifth force. These were very early days in the fifth force effort, but as the Consulting Editor in the AIP Books Division he was interested in such a project irrespective of what the eventual outcome would be. Following an exchange of letters in the ensuing months I received a letter from Rita on May 4, 1988 enclosing a contract. As originally envisioned, this book would be co-authored by Sam Aronson, Carrick Talmadge, and me. I drafted an outline of the proposed book on June 30, 1988, and by August 31, 1988 all three of us had signed and returned the contracts. In a subsequent letter dated November 15, 1989 from Tim Taylor, then the manager for AIP of the division in charge of our book, the target date for completing this book was set at August 1990.

The aforementioned dates are of interest for historical reasons, but primarily because they reveal how much longer it took for us to complete the book than we had anticipated. To start with, Sam was the Deputy Chairman of the Physics Department at Brookhaven at the time we signed the contract, and would eventually become Chairman as we have previously noted. Given his administrative responsibilities, Sam decided that it would be best if he were not a co-author. Carrick and I carried on, but each of us had other research and/or teaching responsibilities which had higher priority. We divided the topics in my earlier outline according to our respective interests and wrote as rapidly as our schedules allowed.

As has been my practice for many years, I broke up my assigned work into individual segments, and began with the segment that was easiest to write. This was on the significance of the shape of B/μ across the periodic table, and I handed my draft to my secretary Nancy Schnepp on January 8, 1991. It began with the words (Fischbach and Talmadge 1999, p. 23): "It is instructive to plot B/μ as a function of atomic number Z for the elements in the periodic table." As indicated by the date on my first segment, we had obviously missed the proposed August 1990 deadline even before we started, and the situation only got worse. Fortunately AIP kept in touch with us, and were extremely understanding.

As time went on my embarrassment continued to increase, and in 1994 an opportunity arose in which I was able to reflect on this in a more public manner. On August 7, 1994 my graduate student Harry Kloor became the first person anywhere to receive two Ph.D. degrees for two completely different projects, in two different areas (physics and chemistry), *on the same day!* Given the novelty of this accomplishment, the *New York Times* sent a photographer to the graduation ceremony, and the *Times* did a story on him on August 8 (p. A6). As the chair of his physics Ph.D. committee, and also a member of his chemistry Ph.D. committee, I was asked to reflect on his achievement, which included defending both theses on the same day. My response was instructive: "What is intimidating is that in four months he wrote these two theses totally more than 700 pages, and I'm struggling to write a book with a co-author and we've barely done 200 pages in several years."

Eventually, however, the book was completed and we sent it off on April 9, 1997 to Maria Taylor who was the editor then in charge of our book. Totaling more than 300 pages, it is an attempt to give a beginning graduate student an introduction to all of the relevant facets of research into the fifth force from both the experimental and theoretical viewpoints, as they were understood by us. By the time our book was actually published in 1999, AIP had joined forces with Springer-Verlag, so that our book appeared as a Springer title.

6.8   Epilogue

As noted in the Introduction, approximately 30 years have elapsed since the publication in PRL of our original paper on the EPF experiment, and so it is appropriate to reflect on what we have learned during this time about a possible fifth force. With the exception of the EPF experiment itself, and possibly the Thieberger floating ball experiment (Thieberger 1987), there is at present no evidence for any deviations from the predictions of Newtonian gravity on any length scale from the solar system down to sub-atomic scales. This conclusion, which applies to both composition-dependent and composition-independent tests, as well as to data on the behavior of the $K^0-\overline{K}^0$ system, is supported by dozens of experiments and hundreds of phenomenological papers.

However, questions remain about the EPF experiment, and to a lesser extent about the Thieberger experiment, and so we cannot close the book on the fifth force story quite yet. Broadly speaking, the EPF correlations could arise from a broad class of interactions characterized by a potential of the form

$$V_{ij} = B_i B_j F(\mathbf{r}_i, \mathbf{r}_j, \mathbf{v}_i, \mathbf{v}_j, \mathbf{s}_i, \mathbf{s}_j, ?) \qquad (6.31)$$

where B_i and B_j are the baryon numbers of the samples, and where $F(\ldots)$ is a function of the other variables (such as position $\mathbf{r}$, velocity $\mathbf{v}$, spin $\mathbf{s}$, etc.) upon which V_{ij} could depend. The critical point in (6.31) is that since B_i and B_j are non-classical quantities, it has not yet been proven possible to account for the EPF correlation in terms of classical systematic effects such as temperature or gravity gradients. Although there may be other systematic effects to be reckoned with, it is clear from what we know that Eötvös, perhaps the greatest "classical" physicist of his time, worried about these in great detail.

It might then be argued that this correlation is just a statistical fluke. However, as noted in Fischbach et al. (1988), the likelihood that EPF obtained $\Delta \kappa \neq 0$ by a statistical accident is extremely small, approximately 5×10^{-12}. Moreover, in a comment at a Moriond conference, de Rújula noted that for the eight "good" points in Figs. 2–5 of Fischbach et al. (1988) the probability of simply getting the *sequence* correct is $2/8! \approx 5 \times 10^{-5}$. Finally, the likelihood of accidentally obtaining approximately the same accelerations for Pt and $CuSO_4 \cdot 5H_2O$, as discussed in Sect. 6.2.3.1, adds to the burden carried by any argument that the EPF data are merely a statistical anomaly.

There is clearly some "tension" between the many careful experiments, most notably from the Eöt-Wash group, which see no evidence for a fifth force, and the EPF experiment. What we can say, however, is that the simple model for a fifth force proportional to baryon number, as presented in our original PRL, is clearly not supported by the totality of existing data. However, we cannot at this stage dismiss the possibility that the function $F(\ldots)$ in (6.31) above could be quite different from what we originally proposed, in such a manner as to admit the possibility of a different kind of fifth force.

Although the final chapters in the fifth story are yet to be written, it is clear that the EPF data have already had a significant impact on gravitational physics by motivating a large number of new (and sometimes novel) experiments and theories. On the experimental side, the torsion balance experiments of the Eöt-Wash group (Adelberger et al. 2009), of Nelson et al. (1988), Boynton (1988), Fitch et al. (1988), and others can be viewed as direct descendants of the EPF experiment, just as that experiment is the descendant of the Guyòt experiment (Fischbach and Talmadge 1999). However, the EPF experiment also stimulated a large number of novel gravitational experiments. These include the floating ball experiments of Thieberger (1987) and Bizzeti et al. (1989); the dropping experiments of Niebauer et al. (1987), Cavasinni et al. (1986), and Kuroda and Mio (1989); the pumped lake experiments of Hipkin and Steinberger (1990) and Cornaz (1994); the Laplacian detector of Moody and Paik (1993); and of course the various tower experiments discussed earlier (Sect. 6.3.8). Finally, the EPF experiment has no doubt played a role in motivating the upcoming MICROSCOPE experiment, which will be the first space-based test of the Weak Equivalence Principle (Touboul and Rodrigues 2001).

On the theoretical side, the early work by Fujii (1971, 1972, 1974, 1975, 1981), Fujii and Nishino (1979), Gibbons and Whiting (1981), and others discussed above, along with the many theories motivated by the EPF data, have drawn attention to the connection between low-energy gravity experiments and high-energy elementary particle physics. This connection, which is explored in Fischbach and Talmadge (1999), can be summarized as follows. Two natural mass scales arise in elementary particle physics, the nucleon mass $m_N \sim 1\,\text{GeV}/c^2$, and the Planck mass $M_P \equiv \sqrt{\hbar c / G_N} \sim 10^{19}\,\text{GeV}/c^2$, where G_N is the Newtonian gravitational constant. Their ratio $m_N/M_P \equiv \sqrt{f^2/\hbar c} \sim 10^{-19}$ defines a new dimensionless constant f which is the analog for some putative new force of the electromagnetic charge e. In many theories the product $\mu \equiv m_N \sqrt{f^2/\hbar c} \sim 10^{-10}\,\text{eV}/c^2$ defines yet another mass scale whose Compton wavelength $\lambda = \hbar/\mu c \approx 2000\,\text{m}$. If μ is the mass of a light bosonic field, then the combination of the parameters f and μ could characterize a new field of gravitational strength whose influence would extend over macroscopic distances. It follows that a search for new macroscopic fields of gravitational strength is yet another means of studying high-energy particle physics. As noted in Sect. 6.6, theories which introduce additional compact spatial dimensions provide yet another link between gravitation and high-energy physics.

In our original PRL we attempted to bring together three anomalies that presented themselves in the 1986 time frame (Fischbach et al. 1986c, Fig. 1; Schwarzschild 1986). These were the EPF data, the discrepancy between the geophysical determinations of G_N and the laboratory value, and the anomalous energy dependence of the K^0–$\overline{K}^0$ parameters, as discussed in Appendix 2. We have already considered the EPF data, and also noted that the original results of Stacey et al. (1987b) were likely due to "terrain bias", as discussed by Bartlett and Tew (1989a). This leaves the puzzling energy dependence of the K^0–$\overline{K}^0$ parameters as the remaining anomaly to be explored.

At the time I arrived at the University of Washington in August 1985, an experiment was underway at Fermilab measuring the mean life τ_S of K_S^0 over the momentum range 100–350 GeV/c. Sam Aronson, Carrick Talmadge, and I were very interested in this experiment for obvious reasons, and through Sam we maintained contact with this group as they analyzed their data. When the results of this experiment were published (Grossman et al. 1987), they revealed no dependence of τ_S on the K_S^0 momentum. Understandably, this had the effect of eliminating the second "leg" of our putative 3-way coincidence among the above anomalies depicted in Fig. 1 of Fischbach et al. (1986c). The experiment of Grossman et al. was done quite carefully, especially given that they were fully aware of the ABCF results, and made repeated references to them.

However, in contrast to the result of Stacey and Tuck, no explanation for the apparently anomalous results obtained by ABCF from Fermilab E621 has emerged. In this way the situation with respect to the ABCF results is somewhat similar to that for the EPF data. With respect to ABCF, Grossman et al. carefully note the difference between their experiment and E621, including the fact that their experiment studied decays from K_S^0 made in proton–tungsten collisions, rather than via K_S^0 regeneration as in E621. Additionally they chose a proper time range where "the contribution of CP non-conservation is insignificant." However, one difference which was not noted is that the E621 beam line was not horizontal (i.e., parallel to the Earth's surface), but rather entered the ground at approximately 8.25×10^{-3} rad to a detector below ground. The possibility that this difference could be relevant has been raised privately with me by Gabriel Chardin, who has independently explored the possibility that CP-violation could be due to some external field (Chardin 1990, 1992). Given that there is no fundamental theory of CP-violation at present, such a mechanism—although unlikely—cannot be excluded at present.

The situation with respect to the ABCF analysis of the E621 data reminds me of a conversation I had some years ago with Melvin Schwartz, who shared the Nobel Prize with Leon Lederman and Jack Steinberger for the discovery of the muon neutrino. I had been invited to talk at Brookhaven on the fifth force, following which several of us went to dinner. In reflecting on the EPF experiment, Schwartz told me of an experiment he tried to carry out some years earlier where he kept getting the "wrong" result. I do not recall why he thought the result was wrong, whether because it disagreed with another experiment or with theory. In any case he kept trying to look for something amiss in his experiment, but to no avail. Finally he decided to disassemble the experiment completely, lead brick by lead brick, and then rebuild it from scratch. For whatever reason, the rebuilt experiment (using exactly the same equipment) now obtained the "right" answer. Schwartz was not able to figure out why these two seemingly identical versions of the same experiment gave different results, and this obviously continued to trouble him.

Although we may never figure out why E621 gave the results obtained by ABCF, I suspect that in time we will eventually understand the EPF data, whatever they reveal in the end. Perhaps there is some subtle detail in E621 or EPF to which we are not paying attention, which is the secret. I am reminded of an appropriate line from the novel *A Taste for Honey* by H.F. Heard (1980):

> This situation is in some way what we all confront in life: those people and events which we treat most contemptuously and thoughtlessly are just those which, watching us through their mask of insignificance, plead with us to understand and feel, and failing to impress and win us, have no choice but to condemn us, for we have already condemned ourselves.

It might thus be an amusing resolution of the fifth force story if the understanding of the EPF experiment was hiding in plain sight all along.

Finally, let me conclude with an update of my co-authors on our PRL. As noted previously, Sam Aronson became chairman of the Physics Department at Brookhaven National Laboratory, and eventually the Director of Brookhaven. He is now (2015) President of the American Physical Society. Carrick Talmadge received his Ph.D. under my supervision in 1987, and eventually switched his interest to acoustics and the human ear. He is now a senior scientist and research associate professor with the National Center for Physical Acoustics at the University of Mississippi. Daniel Sudarsky received his Ph.D. under my supervision in 1989, and is currently a professor at UNAM in Mexico City. Aaron Szafer left Purdue in 1986 with a Master's degree, and received his Ph.D. at Yale in 1990. He is now a technical program manager at the Allen Institute for Brain Science in Seattle.

Acknowledgements I wish to express my deep indebtedness to Dennis Krause whose close collaboration with me on this project has allowed it to be completed in a timely manner. I also wish to thank Allan Franklin for his many helpful editorial suggestions on the early drafts of this manuscript, and Nancy Schnepp for her assistance in organizing this paper. Finally I wish to express my thanks to my co-authors Sam Aronson, Daniel Sudarsky, Aaron Szafer, and Carrick Talmadge who, along with Riley Newman, Peter Thieberger, Mike Mueterthies, and my colleagues at the University of Washington who have helped to clarify various aspects of the historical narrative.

Appendix 1 Fifth Force Phenomenology

In this appendix, I present a summary of the fifth force phenomenology adapted from Fischbach and Talmadge (1999). In the formalism assumed in the original PRL (Fischbach et al. 1986a), the total potential energy $V(r)$ between two interacting samples i and j is the sum of the Newtonian potential $V_N(r)$ and a new fifth force potential $V_5(r)$, viz.,

$$V(r) = V_N(r) + V_5(r) = -\frac{G_\infty m_i m_j}{r} + \frac{f^2 B_i B_j}{r} e^{-r/\lambda}$$
$$= -\frac{G_\infty m_i m_j}{r}\left(1 - \frac{f^2 B_i B_j}{G_\infty m_i m_j} e^{-r/\lambda}\right)$$
$$\equiv -\frac{G_\infty m_i m_j}{r}\left(1 + \alpha_{ij} e^{-r/\lambda}\right), \tag{6.32}$$

where G_∞ is the Newtonian gravitational constant in the limit $r \to \infty$, in which case the contribution from $V_5(r) \to 0$. The functional form of $V_5(r)$ is suggested by models in which this contribution arises from the exchange of an appropriate boson of mass m, and hence $\lambda = \hbar/mc$. B_i and B_j are the respective baryon numbers of i and j, and f is the analog for the putative baryonic force of the electromagnetic charge e. It is conventional to express all masses in terms of the mass of hydrogen $m_H = m(_1H^1) = 1.00782519(8)u$, in which case we write $m_i = \mu_i m_H$, and

$$\alpha_{ij} = -\frac{B_i}{\mu_i}\frac{B_j}{\mu_j}\xi, \tag{6.33}$$

where

$$\xi = \frac{f^2}{G_\infty m_H^2} . \tag{6.34}$$

We note from (6.32), (6.33), and (6.34) that in the presence of $V_5(r)$ the potential energy $V(r)$ depends not only on the masses m_i and m_j, but also on the compositions of the samples via their respective values of B_i/μ_i and B_j/μ_j. As we now show, the accelerations of the two test masses j and k in the presence of a common source i (e.g., the Earth) will depend on the compositions of j and k through the difference $(B_j/\mu_j - B_k/\mu_k)$.

Returning to (6.32) we can calculate the force $F(r) = -\nabla V(r)$:

$$\begin{aligned} F(r) &= -\frac{G_\infty m_i m_j}{r^2}\hat{r}\left[1 + \alpha_{ij}\left(1 + \frac{r}{\lambda}\right)e^{-r/\lambda}\right] \\ &\equiv -\frac{G(r)m_i m_j}{r^2}\hat{r} . \end{aligned} \tag{6.35}$$

In the form of (6.35) the force exerted by m_i on m_j is governed by a "variable Newtonian constant" $G(r)$ which depends not only on r, but also on the compositions of i and j. For experiments carried out over distance scales where $r/\lambda \ll 1$ holds, we can write approximately

$$G(r) \approx G(0) \equiv G_0 = G_\infty(1 + \alpha_{ij}) , \tag{6.36}$$

so that G_0 can be identified with the normal laboratory value. At the other extreme for planetary motion, or for some space-based experiments, where $r/\lambda \gg 1$, $G(r) \approx G(\infty) \equiv G_\infty$. The geophysical experiments of Stacey and Tuck (Stacey 1978; Stacey et al. 1981; Stacey and Tuck 1981), which provided part of the motivation for our reanalysis of the EPF experiment, can be viewed as a determination of the difference between G_0 and $G(r)$ for $\lambda \approx 200$ m.

Returning to (6.35) we see that the presence of the term proportional to α_{ij} leads to two general classes of experiments directed towards searching for a possible fifth force through deviations from the predictions of Newtonian gravity. Broadly speaking these are (a) searches for a *composition dependence* of α_{ij} (also called WEP-violation searches), and (b) searches for an *r-dependence* of $G(r)$. The latter are also referred to as tests of the gravitational inverse-square law, or *composition-independent* tests. Although in principle the term proportional to α_{ij} in (6.35) will generally give rise to both composition-dependent effects and to deviations from the inverse-square law, in practice most experiments have been designed to optimize the search for one or the other effect.

In the preceding discussion we have viewed deviations from the predictions of Newtonian gravity as arising from the presence of a new intermediate-range interaction, as in (6.32). However, similar deviations could also arise from the gravitational interaction itself if gravity did not couple to all contributions to the mass-energy of a test mass with a common universal strength. To this end it is useful to view an atom, and particularly the nucleus, as a "universal soup" of particles in which almost any particle and any interaction (real or virtual) can be present, if only fleetingly. Thus, although we may naively think of an atom as being composed of protons, neutrons, and electrons, in reality part of the mass-energy of an atom arises from virtual e^+e^- pairs, $\pi^\pm$ and π^0 mesons, etc. Hence, if any of the real (p, n, e) or virtual (e^+, e^-, $\pi^\pm$, ...) contributions to the mass-energy of an atom behaved anomalously in a gravitational field, this could produce a non-zero result in a WEP or fifth force experiment. Since these are differential experiments, which compare the forces on two samples, detecting these anomalous behaviors depends on choosing samples for which the anomalous contribution(s) comprise different fractions of the total mass-energy of each sample. Thus by an appropriate choice of pairs of samples one can in principle determine whether the anomalous behavior is due to an external fifth force field coupling to baryon number, isospin, etc., or to a fundamental violation of Lorentz invariance (Fischbach 1965; Fischbach et al. 1985), or to some entirely different mechanism. The observation that a Lorentz non-invariant interaction (LNI) can also show up in WEP experiments is of renewed interest at present in connection with more general searches for LNI effects (Mattingly 2005). Typically an anomalous coupling of gravity to a particular form of energy (e.g., the weak interaction contribution E_W to a nucleus) would give rise to a WEP-violating acceleration difference Δa_{1-2} of two test samples of masses M_1 and M_2 having the form

$$\frac{\Delta a_{1-2}}{a} = \eta_W \left(\frac{E_{W1}}{M_1} - \frac{E_{W2}}{M_2} \right) , \tag{6.37}$$

where η_W is the WEP-violating parameter we are seeking to determine (Fischbach et al. 1985).

In addition to their "universality", another feature of WEP experiments which makes them so interesting is their great sensitivity. Existing laboratory experiments can measure fractional acceleration differences $\Delta a/a$ between samples at roughly the 10^{-13} level, and anticipated space-based experiments such as MICROSCOPE (Touboul and Rodrigues 2001), may push the sensitivity down to 10^{-15}–10^{-16}. At these levels the combination of the universality and sensitivity of WEP experiments makes it interesting to search for various higher-order processes which may be conceptually important, but make relatively small contributions to the mass-energy of a nucleus.

This was the motivation we had in mind in 1995 when I joined with my colleagues Dennis Krause, Carrick Talmadge, and Dubravko Tadić to consider the possibility that an anomalous coupling of neutrinos (ν) and/or antineutrinos ($\bar{\nu}$), to gravity. Neutrinos have been a continuing source of surprises in elementary particle physics starting with their very existence, their role in parity non-conservation, and more recently in flavor oscillations and the solar neutrino problem. As virtual particles the exchange of ν–$\bar{\nu}$ pairs of any flavor gives rise to a 2-body interaction among pairs of nucleons which was first calculated by Feinberg and Sucher in the current–current model (Feinberg and Sucher 1968), and later by Feinberg, Sucher, and Au in the Standard Model (Feinberg et al. 1989). This interaction makes a small contribution to the nuclear binding energy, and hence the question is whether an anomaly in this small contribution could nonetheless be large enough to be detectable in a present or future WEP experiment.

In principle the nuclear binding energy contribution from the exchange of ν–$\bar{\nu}$ pairs, which gives rise to a nucleon–nucleon potential energy $V_{\nu\bar{\nu}}(r)$ proportional to $1/r^5$, could be evaluated for a given nucleus in analogy to the evaluation of the Coulomb contribution $V_C(r)$ which is proportional to $1/r$. However, the contribution from a $1/r^5$ potential would diverge as $r \to 0$ were it not for the nucleon–nucleon hard-core separation, $r_c \approx 0.5$ fm, which sets a lower limit on r. As shown in Fischbach et al. (1995), evaluation of $\langle 1/r^5 \rangle$ over a spherical nucleus for $r_c \neq 0$ can be facilitated by use of techniques from the field of geometric probability. This led to the suggestion that an anomalous coupling of gravity to ν or $\bar{\nu}$ could lead to a WEP violation $\Delta a/a \approx 10^{-17}$. Although this is below the nominal sensitivity of current terrestrial experiments, or of the forthcoming space-based MICROSCOPE experiment, it is possible that a ν–$\bar{\nu}$ anomaly could be larger than the predicted nominal value and hence be detected. Should the MICROSCOPE experiment, or any other experiments, detect a WEP-violating anomaly ($\Delta a_{1-2}/a \neq 0$), then in principle future experiments could determine the underlying mechanism for this violation by studying the dependence of the anomaly on the compositions of various pairs of test samples.

The possibilities of searching for an anomalous coupling of gravity to neutrinos via the 2-body potential $V_{\nu\bar{\nu}}(r)$ eventually led to an analysis of many-body contributions arising from neutrino exchange (Fischbach 1996). Although higher-order long-range forces arising from many-body neutrino exchanges are greatly suppressed, they can also be significantly enhanced in some circumstances due to various combinatoric factors. This had led to the suggestion of a lower bound on neutrino masses, $m_\nu \gtrsim 0.4$ eV/c^2 (Fischbach 1996).

Appendix 2   Phenomenology of the Neutral Kaon System

We briefly review the phenomenology of the neutral kaon system which played an important role in motivating our re-analysis of the EPF experiment. As noted in Sect. 6.1.1, when K^0 and $\overline{K}^0$ are produced by strong interactions they are eigenstates of strangeness S, with eigenvalues $S = +1$ (K^0) or $S = -1$ ($\overline{K}^0$). However, because strangeness is not conserved by the weak interactions which govern kaon decays, the eigenstates of the total Hamiltonian are K_L^0 and K_S^0, which are linear combinations of K^0 and $\overline{K}^0$ given by Aronson et al. (1983a)

$$|K_L^0\rangle = \frac{1}{\sqrt{|p|^2 + |q|^2}} \left(p|K^0\rangle + q|\overline{K}^0\rangle \right) , \tag{6.38}$$

$$|K_S^0\rangle = \frac{1}{\sqrt{|p|^2 + |q|^2}} \left(p|K^0\rangle - q|\overline{K}^0\rangle \right) . \tag{6.39}$$

CP conservation implies that $p = q$, and hence the parameter $\epsilon = 1 - q/p$ is a measure of CP-violation, as are the parameters η_{+-} and η_{00} defined by

$$\eta_{+-} = |\eta_{+-}|e^{i\phi_{+-}} = \frac{A(K_L^0 \to \pi^+\pi^-)}{A(K_S^0 \to \pi^+\pi^-)} , \tag{6.40}$$

$$\eta_{00} = |\eta_{00}|e^{i\phi_{00}} = \frac{A(K_L^0 \to \pi^0\pi^0)}{A(K_S^0 \to \pi^0\pi^0)} . \tag{6.41}$$

Numerically (PDG 2014, p. 944),

$$|\eta_{+-}| = 2.232(11) \times 10^{-3} , \qquad |\eta_{00}| = 2.220(11) \times 10^{-3} ,$$

$$\phi_{+-} = 43.51(5)° , \qquad \phi_{00} = 43.52(5)° , \qquad |\epsilon| = 2.228(11) \times 10^{-3} .$$

As discussed in Sect. 6.1.3, the thread connecting kaon decays and our analysis of the EPF experiment emerged from our analysis of Fermilab data on K_S^0 regeneration. This is the phenomenon in which K_S^0 particles can be regenerated from a pure K_L^0 beam by passing that beam through a target such as hydrogen, carbon, or lead. This phenomenon is interesting since it is a probe of strong interaction models such as Regge pole theory. If we temporarily neglect the effects of CP-violation, then from (6.38) and (6.39) we can write approximately

$$|K_L^0\rangle \approx \frac{1}{\sqrt{2}} \left(|K^0\rangle + |\overline{K}^0\rangle \right) , \tag{6.42}$$

$$|K_S^0\rangle \approx \frac{1}{\sqrt{2}} \left(|K^0\rangle - |\overline{K}^0\rangle \right) , \tag{6.43}$$

and inverting (6.42) and (6.43),

$$|K^0\rangle \approx \frac{1}{\sqrt{2}} \left(|K_L^0\rangle + |K_S^0\rangle \right) , \tag{6.44}$$

$$|\overline{K}^0\rangle \approx \frac{1}{\sqrt{2}} \left(|K_L^0\rangle - |K_S^0\rangle \right) . \tag{6.45}$$

We see that a beam of K^0 produced by a strong interaction process such as $\pi^- + p \rightarrow \Lambda^0 + K^0$, would initially consist of approximately equal amplitudes of K_L^0 and K_S^0. Since K_S^0 decays rapidly ($\tau_S \sim 10^{-10}$ s) compared to K_L^0 ($\tau_L \sim 600\tau_S$), a beam of K^0 produced via the strong interaction will eventually become a pure K_L^0 beam after the initial K_S^0 component decays away. However, a K_S^0 component can be regenerated from a pure K_L^0 beam if that beam is passed through matter, as we now discuss.

Consider the possible strong interactions that can occur when a K_L^0 beam passes through matter. As an example, the $\overline{K}^0$ component of K_L^0 can scatter via $\overline{K}^0 + n \rightarrow \Lambda^0 + \pi^0$, whereas strangeness conservation forbids the analogous process where $\overline{K}^0$ is replaced by K^0. Since similar differences arise as well for virtual processes, it follows that the amplitudes $f_K (\bar{f}_K)$ for the elastic scattering of K^0 ($\overline{K}^0$) on matter are in general unequal. It then follows that if $f_K \neq \bar{f}_K$ the relative admixtures of K^0 and $\overline{K}^0$ in a beam which is initially all K_L^0 will be altered when this beam passes through matter. Specifically,

$$|\psi_{\text{in}}\rangle = |K_L^0\rangle \approx \frac{1}{\sqrt{2}} \left(|K^0\rangle + \overline{K}^0\rangle \right) \longrightarrow |\psi_{\text{out}}\rangle \approx \frac{1}{\sqrt{2}} \left(f_K |K^0\rangle + \bar{f}_K |\overline{K}^0\rangle \right) .$$
$$(6.46)$$

Combining (6.44), (6.45), and (6.46), we can then write

$$|\psi_{\text{out}}\rangle \approx \frac{1}{\sqrt{2}} \left[f_K \left(\frac{|K_L^0\rangle + |K_S^0\rangle}{\sqrt{2}} \right) + \bar{f}_K \left(\frac{|K_L^0\rangle - |K_S^0\rangle}{\sqrt{2}} \right) \right]$$
$$\approx \frac{1}{2} \left[\left(f_K + \bar{f}_K \right) |K_L^0\rangle + \left(f_K - \bar{f}_K \right) |K_S^0\rangle \right] .$$
$$(6.47)$$

The second term in (6.47) thus represents the regenerated K_S^0 component resulting from the incident K_L^0 beam scattering on a target. In the special case of scattering in the forward direction ($\theta = 0$), if $f_K(0) \neq \bar{f}_K(0)$ then the regenerated K_S^0 component will be coherent with the unscattered K_L^0 beam, and interesting interference phenomena can be observed. It is useful to relate the regenerated K_S^0 amplitude to the incident K_L^0 amplitude via a complex parameter ρ defined by Aronson et al. (1983a)

$$|K_S^0\rangle \equiv \rho |K_L^0\rangle .$$
$$(6.48)$$

It can be shown that for a target of length L having N nuclei per unit volume ρ is given by Aronson et al. (1983a)

$$\rho = i\pi N \Lambda_S \alpha (L/\Lambda_S) \left[f_K(0) - \bar{f}_K(0) \right] / k ,$$
$$(6.49)$$

where $\Lambda_S = \beta\gamma\tau_S$ is the mean decay length of K_S^0, k is the wave number of K^0, and

$$\alpha(L/\Lambda_S) = \frac{1 - \exp\left(-\frac{1}{2} + i\Delta m\,\tau_S\right) L/\Lambda_S}{\frac{1}{2} - i\Delta m\,\tau_S}, \tag{6.50}$$

with $\Delta m = m_L - m_S$. The function $\alpha(L/\Lambda_S)$ accounts for the fact that the regenerated K_S^0 is decaying in the target with a characteristic length Λ_S, while also producing a phase change relative to K_L^0 due to the K_L^0–K_S^0 mass difference Δm.

Consider now the time evolution of the coherent K^0 state emerging from a target at $t = 0$. Recalling that this state is a superposition of the initial K_L^0 and the regenerated K_S^0, we can express the initial K^0 state $|\Psi(0)\rangle$ as

$$|\Psi(0)\rangle = |K_L^0\rangle + \rho|K_S^0\rangle, \tag{6.51}$$

where we have temporarily suppressed an overall normalization coefficient. Since both K_S^0 and K_L^0 can decay into $\pi^+\pi^-$ (the latter by virtue of CP-violation), then the net $\pi^+\pi^-$ decay amplitude $\langle\pi^+\pi^-|\Psi(0)\rangle$ is given by the coherent superposition of the two terms in (6.51):

$$\langle\pi^+\pi^-|\Psi(0)\rangle = \langle\pi^+\pi^-|K_L^0\rangle + \rho\langle\pi^+\pi^-|K_S^0\rangle. \tag{6.52}$$

Interestingly, the two amplitudes in (6.52) can be roughly comparable: the suppression of $K_L^0 \rightarrow \pi^+\pi^-$ measured by the CP-violating parameter $|\eta_{+-}|$, can be comparable to the suppression of the CP-allowed $K_S^0 \rightarrow \pi^+\pi^-$ decay due to the smallness of ρ. It follows from (6.52) that the resulting $\pi^+\pi^-$ decay rate arising from N_L incident K_L^0 particles is

$$\frac{dI^{+-}}{dt} = \Gamma\left(K_S^0 \rightarrow \pi^+\pi^-\right) N_L \left\{ |\rho|^2 e^{-t/\tau_S} + |\eta_{+-}|^2 e^{-t/\tau_L} \right. \tag{6.53}$$

$$\left. +2|\rho||\eta_{+-}|\exp\left[-\frac{t}{2}\left(\frac{1}{\tau_S} + \frac{1}{\tau_L}\right)\right]\cos(\Delta m\,t + \phi_\rho - \phi_{+-})\right\}.$$

As noted in Sect. 6.1.3, it follows from (6.53) that the energy dependence of the strong interaction phase $\phi_\rho = \phi_\rho(E)$ can be determined in principle from the time-dependence of the oscillatory factor $\cos[\Delta m\,t + \phi_\rho(E) - \phi_{+-}]$ under the assumption that Δm and ϕ_{+-} are energy-independent fundamental constants.

We conclude this appendix and its relevance to the discussion in Sect. 6.1.3 by elaborating on the anomalous energy dependence of $\phi_\rho(E)$ which eventually led to the suggestion that ϕ_{+-} itself may have been energy-dependent. Returning to (6.48), (6.49), (6.50), and (6.51), we see that ϕ_ρ can be expressed as a sum of three contributions (Aronson et al. 1983a):

$$\phi_\rho = \frac{\pi}{2} + \phi_{\text{geo}} + \phi_{21} \equiv \phi_\rho(E) , \tag{6.54}$$

where the geometric phase ϕ_{geo} and ϕ_{21} are given by

$$\phi_{\text{geo}} = \arg\left[\alpha(L/\Lambda_S)\right] , \tag{6.55}$$

$$\phi_{21} = \arg\left[f_K(0) - \bar{f}_K(0)\right]/k . \tag{6.56}$$

Among the three contributions to $\phi_\rho(E)$ the only quantity whose energy dependence is unknown is ϕ_{21}. Thus a measurement of the energy dependence of the phase

$$\Phi \equiv \phi_\rho - \phi_{+-} = \frac{\pi}{2} + \phi_{\text{geo}} + \phi_{21} - \phi_{+-} \tag{6.57}$$

gives a single constraint on the energy dependence of $(\phi_{21} - \phi_{+-})$.

An extensive discussion of models predicting the energy dependence of $\phi_{21}(E)$ is given in Appendix B of Aronson et al. (1983a), along with a comparison to Fermilab data then available from experiment E621. For regeneration in hydrogen the experimentally determined phase $\phi_{21}^{\text{exp}}(H)$ for kaon momenta in the range $35 \leq p_K \leq 105 \, \text{GeV}/c$ was found to be

$$\phi_{21}^{\text{exp}}(H) = \left[-(139.5 \pm 6.6) + (0.28 \pm 0.09)p_K \right] \deg . \tag{6.58}$$

Over the indicated momentum range this momentum (or energy) dependence would give rise to a phase change in $\phi_{21}^{\text{exp}}(H)$ of $(19.3 \pm 6.3)\,\deg$. By way of comparison, typical theoretical models studied in Aronson et al. (1983a) give $\phi_{21}^{\text{exp}}(H) \lesssim 2°$ over the indicated momentum range (see Fig. 6.1).

As discussed in Sect. 6.1.3, the fact that the combination $(\phi_{21} - \phi_{+-})$ exhibited an energy dependence incompatible with any known model for ϕ_{21}, eventually led us to consider the possibility that ϕ_{+-} itself was energy-dependent. Since such an energy dependence could arise from the coupling of the K^0–$\overline{K}^0$ system to an external hypercharge field, the Fermilab data provided a compelling argument to search for possible new long-range forces, and eventually led to our reanalysis of the EPF experiment.

Appendix 3   Dicke Correspondence

Princeton University   **Department of Physics: Joseph Henry Laboratories**
Jadwin Hall
Post Office Box 708
Princeton, New Jersey 08544

November 20, 1985

```
Professor Ephraim Fischbach
Department of Physics, FM-15
Institute for Nuclear Theory
University of Washington
Seattle, WA 98195
```

Dear Professor Fischbach:

I read your letter of November 6th and your preprint with great interest. Table 1 and Figure 1 apparently show a convincing correlation between ΔK and $\Delta(B/\mu)$. The obvious question concerns the origin of this correlation. Could it be due to some experimental difficulty?

One possibility is a temperature gradient effect giving a torque approximately proportional to the length of a sample. Owing to the <u>absence</u> of a 2-fold symmetry axis, this effect <u>does not</u> disappear when the two sample lengths are equal. One might expect such an effect to vary as $c_1 L_1 - c_2 L_2$ where L_1 and L_2 refer to the lengths of upper and lower weights respective and c_1 and c_2 are regression coefficients to be determined by least squares. I do not recall if the sample lengths are given but they could be assumed to be inversely proportional to the density. In this case the torque would be of the form $\underline{c_1/\rho_1} - c_2/\rho_2$.

If the above fit (with two adjustable parameters) should be worse than the one you show, this could strengthen your argument.

I regret that my reprints for the two papers are gone.

Sincerely,

R. H. Dicke

RHD:mrf

Fig. 6.13 First letter from R.H. Dicke, one of the reviewers of our original PRL (See Fig. 6.19 for his actual report)

The University of Washington

Dept. of Physics, FM-15
Seattle, Washington 98195

Institute for Nuclear Theory
Ephraim Fischbach
Visiting Professor of Physics
(206) 543-2898
Bitnet: ephraim@uwaphast

November 27, 1985

Professor Robert H. Dicke
Department of Physics
P.O. Box 708
Princeton University
Princeton, NJ 08544

Dear Professor Dicke:

Thank you for your letter of November 20. We have taken your suggestion and have fitted the EPF data to the form

$$\kappa_1 - \kappa_2 = a + \frac{b}{\rho_1} - \frac{c}{\rho_2}$$

where a, b, and c are constants to be fitted for. The input data for the fit are given in the accompanying Table, and are plotted in the enclosed graph. Here the contours are obtained by fixing ρ_2 to be that of Cu or Pt, depending on the sample. As you can see, the fit is quite poor ($\chi^2 \approx 28$ for 6 degrees of freedom), especially when compared to the fit presented in our paper. As part of our longer paper we are also checking for other correlations as well.

We very much appreciate your suggestion, and would welcome any additional thoughts that you may have.

Sincerely,

Ephraim Fischbach

EF/jl

Fig. 6.14 Response to the first letter from R.H. Dicke (Fig. 6.13)

Princeton University   **Department of Physics: Joseph Henry Laboratories**
Jadwin Hall
Post Office Box 708
Princeton, New Jersey 08544

June 16, 1986

Professor Ephraim Fischbach
Department of Physics, FM-15
University of Washington
Seattle, WA 98195

Dear Professor Fischbach:

   Thank you for your letter of June 6, 1986.

   You raise an interesting question concerning the constancy of the
thermal effects. But have you been able to find anything definite
concerning the dates the Eötvös experiments were performed; I haven't.
The measurements reported appear to be single sets of data not averages
of many days data. The whole series of measurements might have been
taken in a few weeks. It is normal for an experiment of this type to
require much more time for design, construction and debugging than for
observations. (For our experiment I would guess a ratio of 10:1.)
Another question concerns the position of the apparatus, e.g. where was
the outside wall, etc.?

   Best wishes.

                                        Sincerely,

                                        R. H. Dicke

RHD:mrf   R. H. Dicke

Fig. 6.15 Dicke's second letter following up the letter of Fig. 6.14

The University of Washington

Dept. of Physics, FM-15
Seattle, Washington 98195

Institute for Nuclear Theory
Ephraim Fischbach
Visiting Professor of Physics
(206) 543-2898
Bitnet: ephraim@uwaphast

June 19, 1986

Professor R. H. Dicke
Department of Physics
Princeton University
Princeton, NJ 08544

Dear Professor Dicke:

Thank you very much for your letter of June 16. We have been considering in detail the very questions you asked, aided by the limited information we have. This comes in part from the EPF paper itself (as we discuss below), and in part from correspondence with Professor Jeno Barnothy which we enclose. Dr. Barnothy was a Professor at the Eötvös Institute at the the University of Budapest from roughly 1935 until 1948, as well as being a colleague of Pekár. Also enclosed is a translation of the paper by Eötvös, Pekár, and Fekete (which we may already have sent you, but we're trying to make sure that you have received a copy).

Unfortunately we have no definite record of the exact time frame over which EPF performed their experiments. However, in the first paragraph on page 25 of the translation, they mention that the time between individual measurements of the equilibrium balance position was approximately one hour. Our understanding of the measurement procedure (in particular from the first two paragraphs on page 38 of the translation) is that EPF measured in succession the equilibrium position of the torsion balance with the apparatus aligned respectively in the North, East, South, and West directions. If we call the equilibrium positions for these four alignments respectively, n, e, s, and w, then, as we understand it, EPF obtained their measurements of the equilibrium positions in the following temporal order:

$$n_1, \ e_1, \ s_1, \ w_1, \ n_2, \ e_2, \ s_2, \ w_2, \ ...$$

In other words, our understanding is that EPF did not simply make 114 measurements (for example) of n, rotate the apparatus, make 64 measurements of e, etc., in obtaining their values for v and m for the magnalium-Pt observations. If our understanding is correct, then for just the set of measurements of v and m for the magnalium-Pt datum would have taken at least

$$(114 \times 2 + 64 \times 2 \text{ measurements}) \times 1 \text{ hour/measurement} = 356 \text{ hours} \approx 14 + \text{ days.}$$

Fig. 6.16 Our response to Dicke's second letter (Fig. 6.15)

To Professor R. H. Dicke June 19, 1986 Page 2

If we assume the Method I comparisons for magnalium-Pt and snakewood-Pt were done in parallel (since, as you correctly point out, they used different instruments for these two comparisons), and similarly for Method II, then we estimate that the total run-time required to make the entirety of their observations was roughly 3500 hours, or 140+ days run-time. If we assume that they were able to achieve 40 hours run-time per week, this would imply that the experiment took 88+ weeks (if we don't make the above assumption then the observation time would have been over 4000 hours).

Our present view is that your model is sufficiently promising to warrant more detailed study, which is what we have been doing. For example, we have considered in detail the torque exerted on the pendants by a gentle breeze (arising presumably from horizontal thermal gradients), and have shown both qualitatively and quantitatively that your model could work. We would very much like your model to succeed, in the sense of providing a credible explanation of the EPF data, in the (realistic) chance that no effects are seen in the current experiments. The two questions which we have with this model are regarding the constancy of horizontal thermal effects, which we have already discussed, and the fact that the fits for the Pt data do not work that well. One explanation of this could be that the data are pointing to a double-valued function which B/μ is, but $1/\rho$ is not.

We very much welcome your comments.

Sincerely,

Ephraim Fischbach

Carrick Talmadge

Fig. 6.16 (continued)

JENO M. BARNOTHY
833 LINCOLN STREET
EVANSTON, ILLINOIS 60201
—
312 - 328-5729

Dr.Ephraim Fischbach
Department of Physics
Purdue University
West Lafayette, IN   47907

Dear Ephraim:

Thank you for your letter of June 16. I am glad that you had a
good time in Hungary.
        Your drawing is correctly showing where Eotvos and Renner
have made their measurements with respect to the room. I do not
know of course the exact place in the rooms.In Eotvos's case it
was certainly on one of the "Cleopatra's needles", as we called
these pillars. But I do not know how deep these pillars extended
down in the earth,since this was the only building in the
Institute which had no cellar. Renner's room had a cellar about
10 feet high.I do not know about the new building toward west. At
Eotvos's time the entire institute as you have sketched it was
separated from other buildings by at least 200 feet.
        In 1947 I have published a paper on elementary particles, in
which the proton and the neutron had in addition to their real
mass an imaginary mass of 636 electron masses, which had the
properties of a small electric charge. It could numerically
correctly explain the magnetic moment of the earth and the sun,
and the deviations from the equivalence principle in
Eotvos's experiments, but was in contradiction with Dicke's
observations.

                                                    With best wiches

                                                    Jeno Barnothy

Fig. 6.17 Letter from Barnothy relating to the location of the EPF experiment

Appendix 4   Referee's Reports on Our PRL

THE PHYSICAL REVIEW

—— AND ——

PHYSICAL REVIEW LETTERS

EDITORIAL OFFICES · 1 RESEARCH ROAD
BOX 1000 · RIDGE, NEW YORK 11961
Telephone (516) 924-5533
Telex Number: 971599
Cable Address: PHYSREV RIDGENY

11 December 1985

Dr. Ephraim Fischbach
Physics Department, FM-15
University of Washington
Seattle, WA 98195

Re: Reanalysis of the Eotvos experiment

By: Ephraim Fischbach et al.   LL3052

Dear Dr. Fischbach:

The above manuscript has been reviewed by our referee(s).

On the basis of the resulting report(s), we judge that the paper is not suitable for publication in Physical Review Letters in its present form, but might be made so by appropriate revision. Pertinent criticism extracted from the report(s) is enclosed. While we cannot make a definite commitment, the probable course of action if you choose to resubmit is indicated below.

(X) Acceptance, if the editors can judge that all or most of the criticism has been met.
( ) Return to the original referee(s) for judgement.
( ) Submittal to new referee(s) for judgement.

We are returning your manuscript for revision. Please accompany your resubmittal by a summary of the changes made, and a brief response to any criticisms you have not attempted to meet.

Yours sincerely,

Stanley G. Brown
Editor
Physical Review Letters

Fig. 6.18 PRL's editor report of our paper

May be disclosed to the authors

The results of the statistical fit given in eq. (8), and Fig. 1 are quite convincing. If the authors' interpretation of the fit is correct, the conclusion is of great importance and should be published without question. However, this conclusion is revolutionary and, in my opinion, the authors should briefly discuss, other possible less revolutionary interpretations.

I have already suggested to Professor Fischbach the possibility of a torque induced by a temperature gradient at the Eötvös Balance. He has recently examined this possibility with interesting results. I suggest that his results be very briefly summarized in the article.

Fig. 6.19 Dicke's referee report of our paper

Referee's Report
Physical Review Letters
Manuscript # LL3052
Title: Reanalysis of the Eotvos Experiment
Authors: E. Fischbach *et. al.*

I found the result presented in this paper to be very exciting. I highly recommend the paper for publication in Physical Review Letters.

I have a few comments the authors may wish to consider:

The first comments are regarding the references. Frank Stacey presented a summary of terresterial measurements of G and discussed the form of V(r) (Eq.(1)) in a talk presented at the Workshop on Science Underground, which was published as <u>Science Underground</u>, AIP Conference Proceedings #96, Edited by M. M. Nieto *et. al.* , American Institute of Physics (New York 1983), pp. 285. The authors may wish to augment their references 1 and 2 with Prof. Stacey's talk. A minor typographical detail is that the reference to S. Weinberg on page 6 should be reference #16, not reference #15.

My final comment is regarding the effect of a repulsive force on glactic cluster dynamics. On page 6 the authors raise the question of the effect of the mass of the "hyperphoton" on the "missing mass problem" in cosmology. A related question is what is the effect of the repulsive term, however small it might be, on the morphology of galaxies and the dynamics of clusters of galaxies.

It was a pleasure to review this paper.

Fig. 6.20 Sandberg's referee report of our paper

Appendix 5   Feynman Correspondence

5.1   Los Angeles Times Editorial

The Wonder of It All   *L.A Times*
Jan 15/86

Knowledge, it has been said, is like a circle. What is known is inside, and what is unknown is outside. The larger the diameter of this circle of knowledge, the greater its circumference. And the greater its circumference, the more the circle borders on the unknown. Every time a question is answered, new questions are raised that people didn't even know were questions before. There is no end. Knowledge is infinite and unbounded.

So it should not be surprising that recent reports challenge some basic assumptions of modern physics. Mind you, 20th-Century physics has hardly been a stable body of knowledge in the first place. Physicists have been much better able to gather data than to put it all together in a consistent, coherent theory that can both explain and predict. But progress has been made. Two books were published last year that asserted that physics was on the verge of a complete explanation of the universe.

One of the tenets of physics has been that there are four basic forces in the universe—gravity, electromagnetism and the so-called strong and weak forces of nuclear structure. Prodigious efforts have been made to find a Grand Unified Theory that would demonstrate that all four forces are the same.

Now comes word from a team of physicists led by Ephraim Fischbach of Purdue University that there may be a fifth force in the universe that acts against gravity and causes objects to fall at slightly different rates. This force, which they call hypercharge, would contradict the findings of one of the most famous stories in the history of science: Galileo's dropping cannon balls from the Leaning Tower of Pisa to show that (air resistance aside) all objects fall with the same acceleration regardless of weight or material.

Hypercharge is supposed to be very small and to work only on objects that are fairly close to each other (up to about 600 feet), which would explain why it has not been observed before. We called our friend Richard Feynman, the great theoretical physicist at Caltech, and asked what he thought of this theory. Not much, he said. The new paper by Fischbach and his colleagues is based on experimental data collected in 1909 by Roland von Eotvos. It is not clear, Feynman said, that variations in Eotvos' measurements of gravity result from an unknown fifth force. They could just as easily have been caused by variations in the conditions of Eotvos' experiments. Many more experiments need to be done, he said.

But Feynman had no fear that the existence of a fifth force would damage the structure of physics. Science is a *process* of finding out the truth, he said, and the process is as important as the results. Far from being a stumbling block to a Grand Unified Theory, a fifth force could help scientists refine their ideas and choose among competing models.

In the meantime, it is reasonable to insist on more evidence before rewriting the physics texts.

5.2   *Feynman's Letter to the Los Angeles Times*

CALIFORNIA INSTITUTE OF TECHNOLOGY

CHARLES C. LAURITSEN LABORATORY OF HIGH ENERGY PHYSICS

January 15, 1986

Mr. L. Dembart
Science Writer
Los Angeles Times
Los Angeles, CA 90012

Dear Mr. Dembart:

Thank you for mentioning my name in your editorial. If you
have no intention of writing a longer article, would you please consider
the following letter for the "Letters to the Editor" page?

"You reported in an editorial 'The Wonder of It All' about a
proposal to explain some small irregularities in an old (1909) experiment (by Eötvös) as being due to a new "fifth force." You correctly
said I didn't believe it - but brevity didn't give you a chance to tell
why. Lest your readers get to think that science is decided simply by
opinion of authorities, let me expand here.

If the effects seen in the old Eötvös experiment were due to
the "fifth force" proposed by Prof. Fischbach and his colleagues, with
a range of 600 feet it would have to be so strong that it would have had
effects in other experiments already done. For example, measurements
of gravity force in deep mines agree with expectations to about 1%
(whether this remaining deviation indicates a need for a modification
of Newton's Law of gravitation is a tantalizing question). But the
"fifth force" proposed in the new paper would mean we should have found
a deviation of at least 15%. This calculation is made in their paper
by the authors themselves, (a more careful analysis gives 30%). Although
the authors are aware of this (as confirmed by a telephone conversation)
they call this "surprisingly good agreement," while it, in fact, shows
they cannot be right.

Such new ideas are always fascinating, because physicists wish
to find out how Nature works. Any experiment which deviates from expectations according to known laws commands immediate attention because we
may find something new.

But it is unfortunate that a paper containing within itself its
own disproof should have gotten so much publicity. Probably it is a
result of the authors' over-enthusiasm."

Sincerely,

R. P. Feynman

Richard P. Feynman

5.3   Exchange with Feynman

Jan. 24, 1986

Dear Professor Fischbach;   (My secretary says to be sure to tell you I wrote this, not she. She wouldn't make so many errors!)

Thank you for sending me the papers on the k meson and the latest by Prof. Stacey et al about possible gravity anomalies.

I have been asked by the local paper what I thought of your most recent (Jan. 6) letter to the phys. Rev. Letters. I have had to say that as it stands it is clearly wrong, and that the figures you yourself give prove it wrong. An anamoulous force proportional to hyperon number having a range of 200 meters would have to be 16 times bigger than anything seen in mines to account for the Eotvos results; as you yourself say. There is an additional factor of the ratio of the density of the rock 200 meters below Eotvos' lab relative to the average density of the earth that should go into the formulas. The average of Earth's density is 5.5, and I would guess an average surface density near Eotvos' lab of about half that (granite is 2.7). This makes the discrepency worse, about 35 times what is observed in mines.

Thus I must object to such statements in your paper as "We will demonstrate explicitly that the published data of EPF..strongly support the specific values of the parameters (in Eq 2 where lambda is given as 200-+50 meters)..". Or, " if lambda is in fact on the order of 200 meters details of the local matter distribution ..could lead to improved agreement " when in fact it almost certainly leads to worse agreement, and in no conceivable way could it lead to any reasonable possible agreement (within say a factor of two, the upper limit of tolerance for the size of the mine discrepencies if 200 meters is the range).

But upon studying the data of Stacey I agree that it is possible that agreement could be reached. If there is a layer of density d1 overlying one of density d2, calling s the thickness of d1 in units of lambda (this damn textwriter doesn't handle greek letters or equations, easily) I find, easily enough, that if someone tries to measure the gravity constant relative to the laboratory one, by comparing g, the accelaration of gravity at the surface of d1 to the gravity at the bottom of d1 (ie. at the top of d2) he will get an anamalous value if there is this extra force given by the alpha term. The ratio to the lab constant differs from 1 by alpha times   -1 + f*(1 - exp(-s))/s where f is 1 - d2/(2*d1). We apply it first to data in the Austalian mine, where d1 = d2 (so f = .5) nearly, and several depths from 200m to 1000m were measured. They all give about the same gravity anomally of about + .8 % roughly. This could result if lambda is 200m or less and alpha is - .008, the solution quoted in your paper.

On the other hand if lambda exeeds the greatest depth measued, so s is always small we have another solution. But now alpha must be twice as large, as Stacey notes, .016. At first one might think that this would disagree with the Exxon data in the Gulf of Mexico where perhaps d2 = 2*d1.   This would mean the discrepency in the Gulf would be twice that in the mine, 1.6 %. But it is, pretty much!

If alpha is .016 how big must lambda be if th Eotvos data in your paper is to agree? It comes out 3000m. Is there any data that does not permit this? There is only the comparison of the average gravity on the Earth's surface to the satelite accelarations. This ratio falls below one by directly the a = 5.65 * 10^-6 that you want for Eeotvos multiplied by the ratio of the average density of the first

3000m of the earths surface to the density under the laboratory of
Eotvos. Most of the earth is water, so maybe the average density to
3000m is 1.7 or thereabouts, and guessing 2.7 again for the stuff below
the Eotvos lab we expect a deviation of 3.6 * 10^-6. It is claimed in
a paper that I have not studied yet (Rupp) that only one part per
million is allowed here.   But I don't believe it, for most of my
friends around here doubt that enough good sampling has been done over
the Earth, (and the seas, particularly) to claim that a theoretical
deviation of 4 parts per million is definitely excluded.   But I should
say nothing until I study the evidence more carefully. Stacey does
take it more seriously and will not let himself get much above 1000m on
account of it.   If we do that we can't get enough juice to account for
the Eotvos results.

     Well could the Eotvos results be just a coincidence? How big are
the errors, in fact. Now we are getting (you seem to like this sort of
thing) into the horrible uncertainties of possibly misreading reality
in the configuration of chance errors.   It is very easy to fool
yourself. Of course the statistical errors that Eotvos gives are only
that.   There are always systematic uncertainties and how are we to
determine those? Unfortunately Eotvos never (at first sight) appears
to repeat an experiment, so we can see how well he does.   For example
first he measures magnalium and then compares it to platinum. How much
we wish he would come back to magnalium again to see how consistent he
is!   In fact, his original plan on how to make these comparisons had
to be changed when he realized that the torsion constant of his fibre
had unexpectedly changed by 10%. The result is then calculated
supposing all the shift occured at once, when he changed the metal
samples, but there is no check on that. So this first comparison on
page 34 is quite suspect. He does one more the same way with wood and
platinum (which you didn't use) but changes his method presumably
because the method is suspect. If he doesn't like it, and changes his
method of measurement, should we not also be cautious?

     He continues now by taking a succession of measurements in different
directions to try to average out a possibly slowly drifting fibre. But
all the measurements are taken on one sample before the other is
measured (and he never goes back again to check). And now the question
resolves as to whether ambient gradients of gravitational force in the
lab are constant. (By the way, we have not enough data on the
questioned of small electrical forces and the like..). He makes two
more comparison this way on a single balance, (Cu-Pt and the Ag
reaction before and after) and then improves his technique again. This
time he has two balances parallel and nearly in the same place. To
compare A to B he does it on two balances. First, on balance 1 he has
A, and on balance 2 he has B, and measures both balances. He then
reverses putting B on balance 1 and A on balance 2 and measures again.
Combining all four measurements he proposes to get something least
sensitive to the possibly changing ambiance.

     But look, we have what we want, a repeated experiment!   Think of
just one of the balances, say balance 1. What we have done is change a
sample from A to B on that balance — just the kind of thing we had been
doing before (in the Cu — Pt comparison, for example). It is possible
to calculate with expression (20) just what we would get. But the
same pair was also compared on balance 2, in the opposite order, and we
can make a separate determination of the difference as measured in
balance 2. His final result is the average of these two numbers, but

we can use the separate ones to get an idea of the systematic errors
(at least of the previous measurements which only used one balance). I
have done this. For example, taking Cu-H2O, v1/m1-v1'/m1' (with a
small correction for the difference in delta alpha) is -.005 +- .002,
while v2/m2-v2'/m2' is -.006 +- .0015, very nearly the same.
Systematic errors might be only .001 or less if this were the only
sample we had.

The next, CuSO4-Cu gives -.006 one way and +.001 the other. This
indicates a systematic uncertainty of plus-minus .003. And here (page
43) we notice something unusual happened in this case. The number n1
changed by 23 scale units whereas in no other of theses cases did n1 or
n2 change by more than 2 units.. in fact as time went on the
experimental technique gets better and better, these n's remain more
perfectly equal. But the systematic differences remain, differences in
the final experimental numbers in the third column of your paper being
twice the following numbers (ie the two experimental determinations
from balance 1 and from balance 2 considered separately are the average
in the table plus and minus the following numbers): Asbestos-Cu 0.5
: Salt-Cu 0.5 : Solution-Cu 0.2 : Water - Cu 0.1 : Tallow- Cu
0.7. I take it from this that reasonable estimates of the systematic
uncertainties in Eotvos experiment is about 0.4 (parts per 10^8), twice
the statistical error.

Well, that is the best I can do. It is not sure. But that does
make it a very nerve-wracking thing to believe in the various
deviations when they are just about the size of the errors. On the
other hand, they are all in the right direction. Still it is my best
guess after studying things like this for years that you have had the
bad luck of an apparent fit. It isn't as wonderful as we now believe
looking back. For looking forward, it was too large and we have had to
strain Stacey, and further we have added a new idea to the mine data,
that the effect is proportional to hypercharge. It could have been
just a gravity effect proportional to mass. Maybe we latched on to
hypercharge because the accidents of Eotvos were accidently that
direction. It is hard to say, but I"ll bet you ten to one against it.
I'll bet even more than that, when I remember the Stacey effects are
far from certain, as there are systematic effects comparable to the
"effects" there too.

I look forward to seeing the new kaon analysis. The papers you sent
me (which I have not studied carefully, but there again it looks like
someone is taking statistical errors to be expected errors,
disregarding systematic errors indicated by the lack of consistency of
the whole thing)--the papers you sent me say clearly the effect is not
something like hypercharge that changes sign under charge conjugation.
Yet the negative sign of the alpha, the force between like particles
and the fact that the force between unlike (particle and antiparticle)
is always attraction in quantum field theory, means that the field must
be one that changes sign in charge conjugation.

Well thank you for interesting me in these things. I hope you will
keep me posted. I enclose some remarks attributed to me in an
editorial in the L.A. Times, and a letter I wrote to the opinion page
which they haven't printed. I hope, for your sake that I am wrong..
but, of course, I don't think so. All this is just a personal letter
to you, to show you how much your work interested me. I don't intend
to publish any of this, as far as I know. Good luck.

R. P. Feynman

The University of Washington

Dept. of Physics, FM-15
Seattle, Washington 98195

Institute for Nuclear Theory
Ephraim Fischbach
Visiting Professor of Physics
(206) 543-2898
Bitnet: ephraim@uwaphast

April 14, 1986

Professor Richard P. Feynman
Physics Department
Caltech
Pasadena, CA 91125

Dear Professor Feynman:

I heard from Steve Koonin, who is visiting here, that you are back at Caltech, so I am taking this occasion to reply to your letter of January 24.

1) Concerning the comparison of the Eötvös slope $\Delta\kappa/\Delta(B/\mu)$ and that implied by the geophysical data [Eqs. (9) and (10) of our paper] you have already noted that with the revised Stacey data that better agreement could be achieved. However, since we spoke there has been further work on understanding the Eötvös experiment itself, by us and other workers. The result has been to note that local matter anomalies [buildings, etc.] play a far larger role in these experiments than had herefore been appreciated. When their effects are taken into account, they appear to give the dominant contribution to the Eötvös anomaly $\Delta\kappa$. A simple model of the matter distribution, which we discuss in the enclosed paper, can bring the results of Eqs. (9) and (10) into agreement (to within a factor of 2-4), even if we use the original Stacey data. Thus there no longer appears to be any compelling discrepancy between the Eötvös and geophysical data, and at the same time there is no conflict with the satellite results of Rapp. Unfortunately we did not completely understand this at the time that you and I spoke, and so I did not raise this point then.

2) As to whether the Eötvös results are a coincidence this is, of course, always a possibility. However, in the meantime we have included in our analysis the two data which we previously excluded. One of these (snakewood-Pt) is ` very well established: We obtained several snakewood samples and had two of them chemically analyzed. These yielded virtually identical results, which in turn were quite similar to those one would obtain from other more common woods. The remaining material, tallow, is somewhat more uncertain, and the quoted data point give our best guess (with some horizontal error bar understood). The line has been fitted to all the triangular points. In this connection, the data shown are somewhat different than those we quoted, since we have now gone back to their raw data and recalculated the values of $\Delta\kappa$, without rounding off as EPF did. This slightly changes the quoted results.

To Professor Richard P. Feynman   April 14, 1986   Page 2

I may also add that the resulting line is substantially unchanged if we calculate $\Delta(B/\mu)$ for the combination of the brass vial + contents (which occurs since $(B/\mu)_{\text{brass}} \cong (B/\mu)_{\text{Cu}}$). This point has also been noted independently by a number of other authors.

Given the fact that 9 points now seem to fall along the same line, my feeling is that this probably represents some sort of systematic effect, rather than a statistical fluctuation. We are looking in detail at various systematic effects, including variations of Dicke's "thermal gradient" model which we refer to in the paper. However, at the moment, no model we have looked at thus far seems capable of explaining the indicated correlation. In this connection we would especially welcome suggestions from you.

3) With reference to the kaon data, you have noted correctly that we make of point of saying that a C-odd hypercharge field cannot explain those results. However, that entire analysis explicitly assumes that the γ-dependence comes from a new force which is *long-ranged*, which then naturally leads to a characteristic γ-dependence such as

$$\Delta m = \Delta m_0 [1 + b_\Delta \gamma^2] \tag{1}$$

When the force is of intermediate range, then there arises another dimensional parameter, which is the range λ, and these parameters can enter in combinations such as $(\gamma/\Delta m - \lambda)$. For appropriate λ and γ the contributions from these terms in simple models can be comparable, and in any case give a γ-dependence which is more complicated than that given in Eq. (1) above. For this reason the results of the published analysis cannot be taken over to exclude a hypercharge field of intermediate range. Indeed, in some toy models we have examined, a hypercharge field with the indicated properties does indeed give consistent results. Nonetheless it is much to early to be more definitive on this point at present, but we are continuing our work along these lines.

I hope that I have clarified some of the questions in your letter, and I would very much enjoy hearing from you any additional comments and suggestions that you may have.

Sincerely,

Ephraim Fischbach

Appendix 6   Boynton Spoof

American Institute of Physics, 335 East 45th Street, New York, N.Y. 10017 • TELEPHONE (212) 661-9404
Telex 960983/AMINSTPHYS-NYK

Public Information Division
David A. Kalson, Manager

FOR IMMEDIATE RELEASE

The putative "fifth force" took a major step towards reality today with the publication of a paper by a group at the University of Washington. The paper, entitled "Search for an Intermediate-Range Composition-Dependent Force" by P. Boynton, D. Crosby, D. Ekstrom, and A. Szumilo, in the September 28 issue of Physical Review Letters present decisive evidence that objects of different chemical composition accelerate differently in the gravitational field of the Earth.

The suggestion that objects of different compositions accelerate differently in the field of the Earth, emerged from reanalysis of the classic Eötvös experiment by E. Fischbach and collaborators, who was then at the University of Washington. "When I read Fischbach's analysis, I thought it was complete nonsense", said Boynton, and it was to establish this, that Boynton undertook his experiment. "To my surprise, Fischbach turns out to be correct", said Boynton and he adds "---there is absolutely no doubt that our experiment establishes unequivocally the presence of the fifth force. It appears that Fischbach isn't as crazy as I thought," says Boynton.

The experiment of Boynton and collaborators was carried out at a site near Mt. Index in the Northern Cascade Mountains in Washington. "We convinced the Robbins Manufacturing Company that our experiment was the most important experiment in the world of physics", Boynton explains, "---and on this basis Robbins agreed to drill a hole for us in the side of this mountain." The Robbins Co. went further and arranged for the experimental site to have running water and a comfortably cool ambient temperature. Says Anthony Szumilo, a graduate student who worked with Boynton "...this was the most comfortable experimental environment that I have ever worked in."

Future plans for the Boynton group include repeating their experiment with a new pair of materials, to confirm his earlier results. "We know Newton is wrong" says Boynton, and he adds "----we hope that our next experiment will show that Einstein, Bohr, and Heisenberg were also wrong."

Fig. 6.21 Spoof AIP Press Release sent to Paul Boynton. In reality the hole was drilled by the Robbins Company long before Boynton proposed this experiment, and the "running water" at the site was the result of unwanted drainage from Mount Index which complicated the experiment. The experimental site, rather than having a "comfortably cool ambient temperature", was actually unpleasantly dark, cold, and wet

AMERICAN INSTITUTE OF PHYSICS

335 EAST 45 STREET, NEW YORK, NEW YORK 10017 • TELEPHONE (212) 661-9404
Telex 960983/AMINSTPHYS-NYK

September 18, 1987

Professor Paul Boynton
Physics Dept., FM-15
University of Washington
Seattle, WA 98195

Dear Professor Boynton:

    Enclosed please find the news release that we are distributing to
the media in conjunction with the publication of your paper. As is our
practice, we have paraphrased your remarks to make your work more understandable to the average reader, and we hope this meets with your approval.
I have also sent a copy to Professor Fischbach whom the article mentions.
If you have any questions please feel free to get in touch.

                                      Sincerely,

                                      Judy Chamesh
                                      Publicity

JC:cf

cc/E. Fischbach

Fig. 6.22 Spoof cover letter accompanying the "press release" to Boynton (Fig. 6.21). The letter was signed by my secretary Nancy Schnepp to give it a feminine touch, and the name is completely fictitious

References

Adelberger, E.G., Gundlach, J.H., Heckel, B.R., Hoedl, S., Schlamminger, S.: Torsion balance experiments: a low-energy Frontier of particle physics. Prog. Part. Nucl. Phys. **62**, 102–134 (2009)

Antoniadis, I., Arkani-Hamed, N., Dimopoulos, S., Dvali, G.: New dimensions at a millimeter to a Fermi and superstrings at a TeV. Phys. Lett. B **436**, 257–263 (1998)

Arkani-Hamed, N., Dimopoulos, S., Dvali, G.: Phenomenology, astrophysics, and cosmology of theories with submillimeter dimensions and TeV scale quantum gravity. Phys. Rev. D **59**, 086004 (1999)

Aronson, S.H., Bock, G.J., Cheng, H.-Y., Fischbach, E.: Determination of the fundamental parameters of the K^0–$\overline{K}^0$ system in the energy range 30–110 GeV. Phys. Rev. Lett. **48**, 1306–1309 (1982)

Aronson, S.H., Bock, G.J., Cheng, H.-Y., Fischbach, E.: Energy dependence of the fundamental parameters of the K^0–$\overline{K}^0$ system: experimental analysis. Phys. Rev. D **28**, 476–494 (1983a)

Aronson, S.H., Bock, G.J., Cheng, H.-Y., Fischbach, E.: Energy dependence of the fundamental parameters of the K^0–$\overline{K}^0$ system: theoretical formalism. Phys. Rev. D **28**, 495–523 (1983b)

Aronson, S.H., Cheng, H.-Y., Fischbach, E., Haxton, W.: Experimental signals for hyperphotons. Phys. Rev. Lett. **56**, 1342–1345; **56**, 2334(E) (1986)

Asano, Y., et al.: Search for a rare decay mode $K^+ \rightarrow \pi^+ \nu\bar{\nu}$ and axion. Phys. Lett. B **107**, 159–162 (1981)

Asano, Y., et al.: A new experimental limit on $K^+ \rightarrow \pi^+ \gamma\gamma$. Phys. Lett. B **113**, 195–198 (1982)

Bartlett, D.F., Lögl, S.: Limits on an electromagnetic fifth force. Phys. Rev. Lett. **61**, 2285–2287 (1988)

Bartlett, D.F., Tew, W.L.: Possible effect of the local terrain on the Australian fifth-force measurement. Phys. Rev. D **40**, 673–675 (1989a)

Bartlett, D.F., Tew, W.L.: Possible effect of the local terrain on the North Carolina tower gravity experiment. Phys. Rev. Lett. **63**, 1531 (1989b)

Bartlett, D.F., Tew, W.L.: Terrain and geology near the WTVD tower in North Carolina: implications for non-Newtonian gravity. J. Geophys. Res. **95**, 17363–17369 (1990)

Bell, J.S., Perring, J.K.: 2π decay of the K_2^0 meson. Phys. Rev. Lett. **13**, 348–349 (1964)

Bernstein, J., Cabibbo, N., Lee, T.D.: CP invariance and the 2π decay mode of the K_2^0. Phys. Lett. **12**, 146–148 (1964)

Bizzeti, P.G.: Significance of the Eötvös method for the investigation of intermediate-range forces. Il Nuovo Cimento **94B**, 80–86 (1986)

Bizzeti, P.G., et al.: Search for a composition-dependent fifth force. Phys. Rev. Lett. **62**, 2901–2904 (1989)

Bod, L., Fischbach, E., Marx, G., Náray-Ziegler, M.: One hundred years of the Eötvös experiment. Acta Phys. Hung. **69**, 335–355 (1991)

Bordag, M., Klimchitskaya, G.L., Mohideen, U., Mostepanenko, V.M.: Advances in the Casimir Effect. Clarendon Press, Oxford (2015)

Boslough, J.: Searching for the secrets of gravity. Natl. Geogr. **175**(5), 563–583 (1989)

Bouchiat, C., Iliopoulos, J.: On the possible existence of a light vector meson coupled to the hypercharge current. Phys. Lett. B **169**, 447–449 (1986)

Boynton, P.E., Crosby, D., Ekstrom, P., Szumilo, A.: Search for an intermediate-range composition-dependent force. Phys. Rev. Lett. **59**, 1385–1389 (1987)

Boynton, P.E.: How well do we understand the Torsion balance? In: Fackler, O., Thanh Vân Trân, J. (eds.) 5th Force-Neutrino Physics, Proceedings of the XXIIIrd Rencontre de Moriond (VIIIth Moriond Workshop), pp. 431–444. Editions Frontiéres, Gif-sur-Yvette (1988)

Braginskii, V.B., Panov, V.I.: Verification of the equivalence of inertial and gravitational mass. Sov. Phys. JETP **34**, 463–466 (1972)

Cavasinni, V., Iacopini, E., Polacco, E., Stefanini, G.: Galileo's experiment on free-falling bodies using modern optical techniques. Phys. Lett. A **116**, 157–161 (1986)

Chardin, G.: CP violation: a matter of gravity? In: Thanh Vân Trân, J. (ed.) CP Violation in Particle and Astrophysics, pp. 377–385. Editions Frontiéres, Gif-sur-Yvette (1990)

Chardin, G.: CP violation. A matter of (anti) gravity? Phys. Lett. B **282**, 256–262 (1992)

Chen, Y.-J., et al.: Isoelectronic Measurements Yield Stronger Limits on Hypothetical Yukawa Interactions in the 40–8000 nm range (2014). arXiv:1410.7267v1

Chu, S.Y., Dicke, R.H.: New force or thermal gradient in the Eötvös experiment? Phys. Rev. Lett. **57**, 1823–1824 (1986)

Colella, R., Overhauser, A.W., Werner, S.A.: Observation of gravitationally induced quantum interference. Phys. Rev. Lett. **34**, 1472–1474 (1975)

Cornaz, A., Hubler, B., Kündig, W.: Determination of the gravitational constant G at an effective interaction distance of 112 m. Phys. Rev. Lett. **72**, 1152–1155 (1994)

De Bouard, X., Dekkers, D., Jordan, B., Mermod, R., Willitts, T.R., Winter, K., Scharff, P., Valentin, L., Vivargent, M., Bott-Bodenhausen, M.: Two-pion decay of K_2^0 at 10 GeV/c. Phys. Lett. **15**, 58–61 (1965)

Decca, R.S., et al.: Precise comparison of theory and new experiment for the Casimir force leads to stronger constraints on thermal quantum effects and long-range interactions. Ann. Phys. **318**, 37–80 (2005a)

Decca, R.S., et al.: Constraining new forces in the Casimir regime using the isoelectronic technique. Phys. Rev. Lett. **94**, 240401 (2005b)

Dicke, R.H.: The Eötvös experiment. Sci. Am. **205**(6), 81–94 (1961)

Eckhardt, D.H., et al.: Tower gravity experiment: evidence for non-Newtonian gravity. Phys. Rev. Lett. **60**, 2567–2570 (1988)

Eötvös, R.V., Pekár, D., Fekete, E.: Beiträge zum Gesetze der Proportionalität von Trägheit und Gravität. Annalen der Physik (Leipzig) **68**, 11–66 (1922)

Faller, J.E., Fischbach, E., Fujii, Y., Kuroda, K., Paik, H.J., Speake, C.C.: Precision experiments to search for the fifth force. IEEE Trans. Instrum. Meas. **38**, 180–188 (1989)

Feinberg, G., Sucher, J.: Long-range forces from neutrino-pair exchanges. Phys. Rev. **166**, 1638–1644 (1968)

Feinberg, G., Sucher, J., Au, C.-K.: The dispersion theory of dispersion forces. Phys. Rep. **180**, 83–157 (1989)

Fischbach, E.: Coupling of internal and quantum space-time symmetries. Phys. Rev. **137**, B642–B644 (1965)

Fischbach, E.: Tests of General Relativity at the Quantum Level. In: Bergmann, P.G., De Sabbata, V. (eds.) Cosmology and Gravitation, pp. 359–373. Plenum, New York (1980)

Fischbach, E.: Experimental constraints on new cosmological fields. In: Galić, H., Guberina, B., Tadić, D. (eds.) Phenomenology of Unified Theories, pp. 156–180. World Scientific, Singapore (1984)

Fischbach, E.: Long-range forces and neutrino mass. Ann. Phys. **247**, 213–291 (1996)

Fischbach, E., Freeman, B.S.: Testing general relativity at the quantum level. Gen. Relativ. Gravit. **11**, 377–381 (1979)

Fischbach, E., Freeman, B.S.: Second-order contribution to the gravitational deflection of light. Phys. Rev. D **22**, 2950–2952 (1980b)

Fischbach, E., Nakagawa, N.: Apparatus-dependent contributions to $g-2$ and other phenomena. Phys. Rev. D **30**, 2356–2370 (1984a)

Fischbach, E., Nakagawa, N.: Intrinsic apparatus-dependent effects in high-precision atomic physics experiments. In: Van Dyck, R.S., Jr., Fortson, E.N. (eds.) Ninth Interactional Conference on Atomic Physics: Satellite Workshop and Conference Abstracts, p. 55. World Scientific, Singapore (1984b)

Fischbach, E., Talmadge, C.: The fifth force: an introduction to current research. In: Fackler, O., Thanh Vân Trâh, J. (eds.) 5th Force-Neutrino Physics, Proceedings of the XXIIIrd Rencontre de Moriond (VIIIth Moriond Workshop), pp. 369–382. Editions Frontiéres, Gif-sur-Yvette (1988)

Fischbach, E., Talmadge, C.: Six years of the fifth force. Nature **356**, 207–215 (1992a)

Fischbach, E., Talmadge, C.: Present status of searches for non-Newtonian gravity. In: Sato, H., Nakamura, T. (eds.) Sixth Marcel Grossmann Meeting on Recent Developments in Theoretical and Experimental General Relativity, Gravitation and Relativistic Field Theories, Part B, pp. 1122–1132. World Scientific, Singapore (1992b)

Fischbach, E., Talmadge, C.L.: The Search for Non-Newtonian Gravity. AIP Press/Springer, New York (1999)

Fischbach, E., Freeman, B.S., Cheng, W.-K.: General-relativistic effects in hydrogenic systems. Phys. Rev. D **23**, 2157–2180 (1981)

Fischbach, E., Cheng, H.-Y., Aronson, S.H., Bock, G.J.: Interaction of the K^0–$\overline{K}^0$ system with external fields. Phys. Lett. **116B**, 73–76 (1982)

Fischbach, E., Haugan, M.P., Tadić, D., Cheng, H.-Y.: Lorentz noninvariance and the Eötvös experiments. Phys. Rev. D **32**, 154–162 (1985)
Fischbach, E., Sudarsky, D., Szafer, A., Talmadge, C., Aronson, S.H.: Reanalysis of the Eötvös experiment. Phys. Rev. Lett. **56**, 3–6 (1986a). [Erratum: Physical Review Letters **56**, 1427]
Fischbach, E., Sudarsky, D., Szafer, A., Talmadge, C., Aronson, S.H.: Fischbach et al. respond. Phys. Rev. Lett. **57**, 2869 (1986b)
Fischbach, E., et al.: A new force in nature? In: Geesaman, D.F. (ed.) Intersections Between Particle and Nuclear Physics, AIP Conference Proceedings No. 150, pp. 1102–1118. American Institute of Physics, New York (1986c)
Fischbach, E., Sudarsky, D., Szafer, A., Talmadge, C., Aronson, S.H.: The fifth force. In: Loken, S.C. (ed.) Proceedings of the XXIII International Conference on High Energy Physics, vol. II, pp. 1021–1031. World Scientific, Singapore (1987)
Fischbach, E., Sudarsky, D., Szafer, A., Talmadge, C.: Long-range forces and the Eötvös experiment. Ann. Phys. (New York) **182**, 1–89 (1988)
Fischbach, E., Talmadge, C.: Ten years of the fifth force. In: Ansari, R., Giruad-Héraud, U., Van Tran Thanh, J. (eds.) Dark Matter in Cosmology, Quantum Measurements, Experimental Gravitation. Proceedings of the 31st Rencontres de Moriond (16th Moriond Workshop), pp. 443–451. Editions Frontiéres, Gif-sur-yvette (1996)
Fischbach, E., Gillies, G.T., Krause, D.E., Schwan, J.G., Talmadge, C.: Non-Newtonian gravity and new weak forces: an index of measurements and theory. Metrologia **29**, 213–260 (1992)
Fischbach, E., et al.: New geomagnetic limit on the photon mass and on long-range forces coexisting with electromagnetism. Phys. Rev. Lett. **73**, 514–519 (1994)
Fischbach, E., Krause, D.E., Talmadge, C., Tadić, D.: Higher-order weak interactions and the equivalence principle. Phys. Rev. D **52**, 5417–5427 (1995)
Fischbach, E., et al.: Testing gravity in space and at ultrashort distances. Class. Quantum Gravity **18**, 2427–2434 (2001)
Fischbach, E., Krause, D.E., Decca, R.S., López, D.: Testing Newtonian gravity at the nanometer distance scale using the iso-electronic effect. Phys. Lett. A **318**, 165–171 (2003)
Fitch, V.L., Isaila, M.V., Palmer, M.A.: Limits on the existence of a material-dependent intermediate-range force. Phys. Rev. Lett. **60**, 1801–1804 (1988)
Floratos, E.G., Leontaris, G.K.: Low scale unification, Newton's law and extra dimensions. Phys. Lett. B **465**, 95–100 (1999)
Franklin, A.: The Rise and Fall of the Fifth Force. American Institute of Physics, New York (1993)
Fujii, Y.: Dilaton and possible non-Newtonian gravity. Nature (Phys. Sci.) **234**, 5–7 (1971)
Fujii, Y.: Scale invariance and gravity of hadrons. Ann. Phys. (New York) **69**, 494–521 (1972)
Fujii, Y.: Scalar–tensor theory of gravitation and spontaneous breakdown of scale invariance. Phys. Rev. D **9**, 874–876 (1974)
Fujii, Y.: Spontaneously broken scale invariance and gravitation. Gen. Relativ. Gravit. **6**, 29–34 (1975)
Fujii, Y.: Composition independence of the possible finite-range gravitational force. Gen. Relativ. Gravit. **13**, 1147–1155 (1981)
Fujii, Y., Nishino, H.: Some phenomenological consequences of the super higgs effect. Zeitschriji für Physik C, Particles and Fields **2**, 247–252 (1979)
Galbraith, W., Manning, G., Taylor, A.E., Jones, B.D., Malos, J., Astbury, A., Lipman, N.H., Walker, T.G.: Two-pion decay of the K_2^0 meson. Phys. Rev. Lett. **14**, 383–386 (1965)
Gibbons, G.W., Whiting, B.F.: Newtonian gravity measurements impose constraints on unification theories. Nature **291**, 636–638 (1981)
Grossman, N., et al.: Measurement of the lifetime of K_S^0 mesons in the momentum range 100 to 350 GeV/c. Phys. Rev. Lett. **59**, 18–21 (1987)
Hall, A.M., Armbruster, H., Fischbach, E., Talmadge, C.: Is the Eötvös experiment sensitive to spin? In: Hwang, W.-Y.P., et al. (eds.) Progress in High Energy Physics. Proceedings of the Second International Conference and Spring School on Medium and High Energy Nuclear Physics, pp. 325–339. North Holland, New York (1991)
Heard, H.: The Amazing Mycroft Mysteries, pp. 196–197. Vanguard Press, New York (1980)

Hipkin, R.G., Steinberger, B.: Testing Newton's law in the Megget water reservoir. In: Rummel, R., Hipkin, R.G. (eds.) Gravity, Gradiometry, and Gravimetry, Symposium No. 103, pp. 31–39. Springer, New York (1990)

Holding, S.C., Tuck, G.J.: A new mine determination of the Newtonian gravitational constant. Nature **307**, 714–716 (1984)

Hólmansson, S., Sanders, C., Tucker, J.: Concise Icelandic–English Dictionary, vol. 294. IDUNN, Reykjavík (1989)

Hoskins, J.K., Newman, R.D., Spero, R., Schultz, J.: Experimental tests of the gravitational inverse-square law for mass separations from 2 to 105 cm. Phys. Rev. D **32**, 3084–3095 (1985)

Hughes, R., Bianconi, P.: The Complete Paintings of Bruegel. Harry N. Abrams, New York (1967)

Jekeli, C., Eckhardt, D.H., Romaides, A.J.: Tower gravity experiment: no evidence for non-Newtonian gravity. Phys. Rev. Lett. **64**, 1204–1206 (1990)

Kammeraad, J., et al.: New results from Nevada: a test of Newton's law using the BREN tower and a high density ground gravity survey. In: Fackler, O., Thanh Vân Trân, J. (eds.) New and Exotic Phenomena '90, Proceedings of the XXVth Rencontre de Moriond, pp. 245–254. Editions Frontiéres, Gif-sur-Yvette (1990)

Kehagias, A., Sfetsos, K.: Deviations from the $1/r^2$ Newton law due to extra dimensions. Phys. Lett. B **472**, 39–44 (2000)

Kloor, H., Fischbach, E., Talmadge, C., Greene, G.L.: Limits on new forces co-existing with electromagnetism. Phys. Rev. D **49**, 2098–2114 (1994)

Krause, W.: A letter from Eötvös. Zeitschrift für Naturforschung **43a**, 509–510 (1988)

Krause, D.E., Fischbach, E.: Isotopic dependence of the Casimir force. Phys. Rev. Lett. **89**, 190406 (2002)

Krause, D.E., Kloor, H.T., Fischbach, E.: Multipole radiation from massive fields: application to binary pulsar systems. Phys. Rev. D **49**, 6892–6906 (1994)

Krauss, L.M.: A fifth farce. Phys. Today **61**(10), 53–55 (2008)

Kreuzer, L.B.: Experimental measurement of the equivalence of active and passive gravitational mass. Phys. Rev. **169**, 1007–1012 (1968)

Kuroda, K., Mio, N.: Galilean test of the composition-dependent force. In: Blair, D.G., Buckingham, M.J. (eds.) Proceedings for the Fifth Marcel Grossmann Meeting on General Relativity, pp. 1569–1572. World Scientific, Singapore (1989)

Lee, T.D., Wu, C.S.: Weak interactions (second section) Chapter 9: decays of neutral K mesons. Ann. Rev. Nucl. Sci. **16**, 511–590 (1966)

Lee, T.D., Yang, C.N.: Conservation of heavy particles and generalized Gauge transformations. Phys. Rev. **98**, 1501 (1955)

Long, D.R.: Experimental examination of the gravitational inverse square law. Nature **260**, 417–418 (1976)

Long, D.R.: Vacuum polarization and non-Newtonian gravitation. Il Nuovo Cimento **55B**, 252–256 (1980)

Lusignoli, M., Pugliese, A.: Hyperphotons and K-meson decays. Phys. Lett. B **171**, 468–470 (1986)

Maddox, J.: Newtonian gravitation corrected. Nature **319**, 173 (1986a)

Maddox, J.: Looking for gravitational errors. Nature **322**, 109 (1986b)

Maddox, J.: Prospects for fifth force fade. Nature **329**, 283 (1987)

Maddox, J.: Making the Geoid respectable again. Nature **332**, 301 (1988a)

Maddox, J.: Reticence and the upper limit. Nature **333**, 295 (1988b)

Maddox, J.: The stimulation of the fifth force. Nature **335**, 393 (1988c)

Maddox, J.: Weak equivalence in the balance. Nature **350**, 187 (1991)

Mattingly, D.: Modern tests of Lorentz invariance. Living Rev. Relativ. **8**, 5 (2005)

Milgrom, M.: On the use of Eötvös-type experiments to detect medium-range forces. Nucl. Phys. **B277**, 509–512 (1986)

Moody, M.V., Paik, H.J.: Gauss's law test of gravity at short range. Phys. Rev. Lett. **70**, 1195–1198 (1993)

Nelson, P.G., Graham, D.M., Newman, R.D.: A 'Fifth Force' search using a controlled local mass. In: Fackler, O., Thanh Vân Trân, J. (eds.) 5th Force-Neutrino Physics, Proceedings of the XXIIIrd Rencontre de Moriond (VIIIth Moriond Workshop), pp. 471–480. Editions Frontiéres, Gif-sur-Yvette (1988)

Neufeld, D.A.: Upper limit on any intermediate-range force associated with Baryon number. Phys. Rev. Lett. **56**, 2344–2346 (1986)

Niebauer, T.M., McHugh, M.P., Faller, J.E.: Galilean test for the fifth force. Phys. Rev. Lett. **59**, 609–612 (1987)

Olive, K.A., et al. (Particle Data Group): Review of particle physics. Chin. Phys. C **38**(9), 1–1676 (2014)

Particle Data Group: Review of particle physics. Eur. Phys. J. C **3**, 1–794 (1998)

Randall, L., Sundrum, R.: Large mass hierarchy from a small extra dimension. Phys. Rev. Lett. **83**, 3370–3373 (1999)

Renner, J.: Kísérleti vizsgálatok a tömegvonzás és a tehetetlenség arányosságáról. Matematikai és Természettudományi Értesitö **53**, 542–568 (1935)

Roll, P.G., Krotkov, R.V., Dicke, R.H.: The equivalence of inertial and passive gravitational mass. Ann. Phys. (N. Y.) **26**, 442–517 (1964)

Romaides, A.J., et al.: Second tower experiment: further evidence for Newtonian gravity. Phys. Rev. D **50**, 3613–3617 (1994)

Romaides, A.J., Sands, R.W., Fischbach, E., Talmadge, C.: Final results from the WABG tower gravity experiment. Phys. Rev. D **55**, 4532–4536 (1997)

Schwarzschild, B.: Reanalysis of old Eötvös data suggests 5th force ... to some. Phys. Today **39**(12), 17–20 (1986)

Simpson, W.M.R., Leonhardt, U. (eds.): Forces of the Quantum Vacuum: An Introduction to Casimir Physics. World Scientific, Singapore (2015)

Speake, C.C., et al.: Test of the inverse-square law of gravitation using the 300 m tower at Erie, Colorado. Phys. Rev. Lett. **65**, 1967–1971 (1990)

Spero, R., et al.: Test of the gravitational inverse-square law at laboratory distances. Phys. Rev. Lett. **44**, 1645–1648 (1980)

Stacey, F.D.: Possibility of a geophysical determination of the Newtonian gravitational constant. Geophys. Res. Lett. **5**, 377–378 (1978)

Stacey, F.D.: Subterranean gravity and other deep hole geophysics. In: Nieto, M.M., et al. (eds.) Science Underground. AIP Conference Proceedings, No. 96, pp. 285–297. American Institute of Physics, New York (1983)

Stacey, F.D.: Gravity. Sci. Prog. **69**(273), 1–17 (1984)

Stacey, F.D.: Gravity: a possible refinement of Newton's law. In: Scott, A. (ed.) Frontiers of Science, pp. 157–170. Blackwell, Oxford (1990)

Stacey, F.D., Tuck, G.J.: Geophysical evidence for non-Newtonian gravity. Nature **292**, 230–232 (1981)

Stacey, F.D., Tuck, G.J.: Non-Newtonian gravity: geophysical evidence. In: Taylor, B.N., Phillips, W.D. (eds.) Precision Measurement and Fundamental Constants II. National Bureau of Standards Special Publication, 617, pp. 597–600. U.S. National Bureau of Standards, Washington (1984)

Stacey, F.D., Tuck, G.J.: Is gravity as simple as we thought? Phys. World **1**(3), 29–32 (1988)

Stacey, F.D., Tuck, G.J., Holding, S.C., Maher, A.R., Moms, D.: Constraint on the planetary scale value of the Newtonian gravitational constant from the gravity profile within a mine. Phys. Rev. D **23**, 1683–1692 (1981)

Stacey, F.D., Tuck, G.J., Holding, S.C., Moore, G.I., Goodwin, B.D., Ran, Z.: Large scale tests of the inverse square law. In: MacCallum, M., et al. (eds.) Abstracts of Contributed Papers, 11th International Conference on General Relativity and Gravitation, Stockholm, p. 627. International Society on General Relativity and Gravitation (1986)

Stacey, F.D., Tuck, G.J., Moore, G.I.: Geophysical tests of the inverse square law of gravity. In: Fackler, O., Thanh Vân Trân, J. (eds.) New and Exotic Phenomena, Proceedings of the XXIInd Rencontre de Moriond, pp. 557–565. Editions Frontiéres, Gif-sur-Yvette (1987a)

Stacey, F.D., Tuck, G.J., Moore, G.I., Holding, S.C., Goodwin, B.D., Zhou, R.: Geophysics and the law of gravity. Rev. Mod. Phys. **59**, 157–174 (1987b)

Stacey, F.D., Tuck, G.J., Moore, G.I.: Quantum gravity: observational constraints on a pair of Yukawa terms. Phys Rev. D **36**, 2374–2380 (1987c)

Stacey, F.D., Tuck, G.J., Moore, G.I.: Geophysical considerations in the fifth force controversy. J. Geophys. Res. **93**, 10575–10587 (1988)

Suzuki, M.: Bound on the mass and coupling of the hyperphoton by particle physics. Phys. Rev. Lett. **56**, 1339–1341 (1986)

Szabó, Z. (ed.): Three Fundamental Papers of Loránd Eötvös. Loránd Eötvös Geophysical Institute of Hungary, Budapest (1998)

Talmadge, C., Aronson, S.H., Fischbach, E.: Effects of local mass anomalies in Eötvös-type experiments. In: Thanh Vân Trân, J. (ed.) Progress in Electroweak Interactions, vol. 1, pp. 229–240. Editions Frontiéres, Gif-sur-Yvette (1986)

Talmadge, C., Berthias, J.-P., Hellings, R.W., Standish, E.M.: Model-independent constraints on possible modifications of Newtonian gravity. Phys. Rev. Lett. **61**, 159–1162 (1988)

Thieberger, P.: Hypercharge fields and Eötvös-type experiments. Phys. Rev. Lett. **56**, 2347–2349 (1986)

Thieberger, P.: Search for a substance-dependent force with a new differential accelerometer. Phys. Rev. Lett. **58**, 1066–1069 (1987)

Thodberg, H.H.: Comment on the sign in the reanalysis of the Eötvös experiment. Phys. Rev. Lett. **56**, 2423 (1986)

Thomas, J., et al.: Testing the inverse-square law of gravity on a 465 m tower. Phys. Rev. Lett. **63**, 1902–1905 (1989)

Touboul, P., Rodrigues, M.: The MICROSCOPE space mission. Class. Quantum Gravity **18**, 2487–2498 (2001)

Weinberg, S.: Do hyperphotons exist? Phys. Rev. Lett. **13**, 495–497 (1964)

Will, C.M.: Theory and Experiment in Gravitational Physics, Rev. edn. Cambridge University Press, New York (1993)

Further Reading

Adelberger, E.G., et al.: Constraints on composition-dependent interactions from the Eöt-Wash experiment. In: Fackler, O., Tran Thanh Van, J. (eds.) (1988), pp. 445–456 (1998)

Eckhardt, D.H., et al.: The North Carolina tower gravity experiment: a null result. In: Fackler, O., Tran Thanh Van, J. (eds.), pp. 237–244 (1990)

Fackler, O., Tran Thanh Van, J. (eds.): New and Exotic Phenomena: Seventh Moriond Workshop. Editions Frontières, Gif sur Yvette (1987)

Fackler O., Tran Thanh Van, J. (eds.): 5th Force Neutrino Physics: Eighth Moriond Workshop. Editions Frontières, Gif sur Yvette (1988)

Fackler, O., Tran Thanh Van, J. (eds.): Tests of Fundamental Laws in Physics: Ninth Moriond Workshop. Editions Frontières, Gif sur Yvette (1989)

Fischbach, E.: The Fifth Force: an introduction to current research. In: Fackler, O., Tran Thanh Van, J. (eds.), pp. 369–382 (1988b)

Fischbach, E., Talmadge, C.: Recent developments in the Fifth Force. Mod. Phys. Lett. A **4**, 2303–2315 (1989)

Luther, G.B., Towler, W.L.: Redetermination of the Newtonian gravitational constant G. Phys. Rev. Lett. **48**, 121–123 (1982)

Meszaros, A.: Did Eötvös Torsion experiment detect Cartan's contortion. Astrophys. Space Sci. **125**, 405–410 (1986)

Overhauser, A.W., Colella, R.: Experimental test of gravitationally induced quantum interference. Phys. Rev. Lett. **33**, 1237–1239 (1974)

Stacey, F.D.: Subterranean gravity and other deep hole geophysics. In: Nieto, M.M., et al. (eds.) Science Underground, pp. 285–297. American Institute of Physics, New York (1983)

Tran Thanh Van, J. (ed.): Progress in Electroweak Interactions, Proceedings of the Lepton Session of the Twenty-First Rencontre de Moriond, pp. 229–240. Editions Frontières, Gif sur Yvette (1986)

© Springer International Publishing Switzerland 2016

A. Franklin, E. Fischbach, *The Rise and Fall of the Fifth Force*,

DOI 10.1007/978-3-319-28412-5

Index

A

ABCF papers   148, 186, 206, 207
Adelberger, E.G.   46, 47, 51, 63, 64, 67,
        69–71, 73, 80–86, 115, 123, 126,
        136, 139–141, 198, 200, 201
AFGL collaboration   188
Air Force Geophysical Laboratory   188
Ander, M.E.   67, 73
Anderson, P.   102–106
Angular acceleration measurements   119
Anti-gravity   176, 182, 195
Antoniadis, I.   201
Arkani-Hamed, N.   201
Armbruster, H.   172, 173
Aronson, S.H.   ix, x, 4–8, 12, 23, 26, 30–40,
        147, 148, 152, 154–157, 182, 184,
        185, 192, 198, 200, 203, 206, 207,
        212, 213, 215
Astronomical gravitation measurements   11
Au, C.-K.   211
Avron, Y.   28

B

Barnothy, J.   186
Barr, S.M.   28
Bars, I.   28, 29, 61
Bartlett, D.F.   68, 71, 72, 75, 80–83, 117, 188,
        191, 192, 206
Baryon coupling   ix, 9, 15, 16, 64, 70, 71, 76,
        118
Baryon number   15, 16, 140, 159, 171, 208

Bayes theorem   102–105
BCL experiment   157, 158
Belinfante, F.   154
Bell, J.S.   156, 157
Bennett, W.R.   68, 69, 117, 127, 136, 137
Bernstein, J.   156, 157
Bertolami, O.   27
Bessel experiment   160
Bizzeti, P.G.   23, 27, 32, 60, 61, 63–65, 71,
        77, 94, 177, 192, 205
Bock, G.J.   5, 148, 154, 155
Bod, L.   182, 186
Boehm, F.   193
Bordag, M.   202
Boslough, J.   187, 198, 199
Bouchiat, C.   26, 185
Boudreaux, A.   123, 127
Boynton, P.E.   52–58, 64, 66, 68, 72, 76, 77,
        80, 81, 83, 115, 124, 200, 201, 232
BP experiment   148–150, 158, 159, 166, 170,
        198
Braginskii, V.B.   12, 148, 159, 166, 168, 170
Brans–Dicke theory   3, 8
Bruegel, P.   156
Buck, P.   150, 166

C

Cabibbo, N.   156, 157
Carosi, R.   8
Carusotto, S.   118, 119, 123, 127
Casimir effect   148, 202, 203
Cavasinni, V.   118, 119, 123, 127
Chen, Y.J.   202, 203

© Springer International Publishing Switzerland 2016
A. Franklin, E. Fischbach, *The Rise and Fall of the Fifth Force*,
DOI 10.1007/978-3-319-28412-5

Y

Z

Printed in the United States
By Bookmasters